A parents' guide to school mathematics

A parents' guide to school mathematics

ALAN TAMMADGE
Headmaster of Sevenoaks School

PHYLLIS STARR
Head of the Mathematics Department
Henrietta Barnett School, London

Cartoons by David and Mark Ingram

CAMBRIDGE UNIVERSITY PRESS
Cambridge
London · New York · Melbourne

Published by the Syndics of the Cambridge University Press
The Pitt Building, Trumpington Street, Cambridge CB2 1RP
Bentley House, 200 Euston Road, London NW1 2DB
32 East 57th Street, New York, NY 10022, USA
296 Beaconsfield Parade, Middle Park, Melbourne 3206, Australia

© Cambridge University Press 1977

First published 1977

Printed in Great Britain at the
University Printing House,Cambridge CB2 1RP

Library of Congress Cataloguing in Publication Data
Tammadge, Alan.
A parents' guide to school mathematics.
(School Mathematics Project handbooks)
Includes index.
1. Mathematics – Study and teaching – Great Britain.
I. Starr, Phyllis, joint author. II. Title.
III. Series: School Mathematics Project. Handbooks.
QA14.G7T35   510'.7   75-46135
ISBN 0 521 21108 5

Contents

Preface

This book was first written several years ago when mathematics curriculum development was at its height. It then lay dormant while I found myself more concerned with school administration. However, I kept having guilty feelings that parents needed a book which would explain something of what we had done and still had in mind to do. It was therefore a great day when Phyllis Starr agreed to read the material, do some radical pruning and work with me in bringing it up to date. I managed to interest David Ingram, then a pupil at my school in producing funny drawings and, when he left, his father Mark took over. These kept our spirits up, a necessary business, for revising is much less fun than writing the original.

Phyllis Starr joins me in thanking the many people who have helped us. In particular Mrs Betty Young who typed the first draft, Mrs Rose Greaves who typed the revised one, Jim Hornsby of Sevenoaks School who took the photographs and all the SMP-ites who read and criticised the work in its various stages. Finally we would like to thank Tony May, our first non-mathematical parent reader, for most useful comments, and the children of his school (at that time Seal Church of England Primary School), as well as our own two schools, for their (sometimes unconscious) contributions.

Alan Tammadge

Introduction

Has there been a mathematical revolution in your child's school? If you are going to read this book then probably there has. The first signs are that your child comes home from school and says, 'Do you know that $2 + 3 = 10$?' Later you may be asked what you know about identity and inverse, the inter-quartile range or the empty set. The revolution is affecting both primary and secondary teaching and although the movement is little more than fifteen years old in the United Kingdom, it shows every sign of being here to stay. It is easier to see this happening than to give satisfactory reasons. Perhaps the primary reason was just dissatisfaction with the level of numeracy in the population at large, aggravated in recent years by the worsening position in the supply of teachers of mathematics. Another was the realisation by educators that mathematics is not a static body of knowledge. Until very recently it would have been possible for a teacher from the nineteenth century to walk straight into your child's classroom, pick up the nearest book and start teaching. Indeed, the book itself might have been familiar! Of course, it is not self-evident that because the syllabus is different, it is necessarily better, but in the course of this book we are going to try to explain why we think the new syllabuses are better, and why we give, as the third good reason, that the new mathematics is − fun.

It will not be possible to tell you about this new mathematics without actually doing some, and we hope and expect that you will be prepared, from time to time, to take pencil and paper and work out a few simple problems. The answers are all at the back of the book.

One of the difficulties in talking about mathematics is to decide whether it is 'pure', that is to say an intellectual treat, an abstract study, an edifice of pure thought, or whether it is 'applied', that is to say a means of getting answers to problems. Most teachers of mathematics in this country would answer that it is both; but this is not the answer that all professional mathematicians would give, nor is it the answer that has been given by teachers of school mathematics in all foreign countries. This is a complicated question, however, best dealt with in a new chapter.

Human pictogram (see p. 38)

1 The place of mathematics in a general education

The purpose of education

The purpose of all education is, we assert, the development of the personality. This is, of course, a cliché and in attempting to see what lies behind it we shall have to disentangle two points of view, the personal and the social. The personal aspect of

'the development of the personality'

education is often related to rewards of one sort or another. The reward may be immediate, a kind word from the teacher, a pat on the head, the glow of virtue at having finished a difficult exercise, or it may be long term in that the thing that has been mastered enables one to shine in company, to carry out one's job better, to tackle a problem that otherwise one would not be able to tackle. The educated person has mental resources; we hear of prisoners in solitary confinement who save themselves from going mad by working long problems in arithmetic or by doing imaginary geometry with the cracks in the ceiling. The goal of education is mastery. There are good psychological reasons why books are divided into chapters and a school career into terms, for a learner needs sign-posts; points at which one can look back and say, 'Thus far have I gone, now for the next step.' A defect of the old mathematics was that the school-leaver could rarely look back with any confidence, and say what his mathematics amounted to. He knew some facts, some techniques, some tricks, but had no answer to the fundamental question, 'What *is* mathematics?'

Socially speaking, it is important to learn things which enable one to live in harmony with one's neighbours and to make a useful contribution to the society in which one lives. At the lowest level we find the need to cope with the arithmetic

[3]

involved in ordinary living. Despite the advent of machines it is obviously still necessary to master the processes of arithmetic in order to be able to add bills or work out change. It is also necessary to know the size of the units met with in everyday life: the mile, the kilometre, the pound, the kilogram, and, particularly at a time of change-over, the relation between pint and litre, and so on. The housewife uses her knowledge of the size of units every day, although the advent of pre-packaged items takes some of the need away. She knows that a box weighing 10 kg will be hard to lift, that 20 grams of pepper would be too much for any domestic dish and that a litre of milk will not go into a teacup. Elementary calculations are also important; it is necessary to calculate the time to cook the Sunday joint or the length of wallpaper required to redecorate a room. In the case of these last two problems, however, it is noteworthy that what is required is an approximation and that the traditional arithmetic practised in schools is not entirely appropriate; to this we will return later.

'Elementary calculations . . . to calculate the time to cook the Sunday joint'

It is also of great social importance in a technological age that there should be an adequate supply of citizens who can cope with more advanced mathematics, so that blocks of flats do not fall down, bearings in engines do not overheat, airports are situated in the proper places and so on. Later we will argue that more and more people are being involved in these more complicated types of calculation and that it is not enough to hope piously that a sufficiency of 'experts' will automatically be produced by the education system.

Numbers are interesting

The previous paragraphs may make it sound as though learning mathematics is rather a solemn and weighty process. In fact, many quite small children find an immediate interest in numbers; they take pleasure in learning to count. At first, they may just like to recite the numerals – '1, 2, 3, 4, 5, once I caught a fish alive'. Later they will count stairs, lamp-posts, their bricks as they pile them, and they will frequently attach an almost mystical significance to their age, and may share with some adults an interest in lucky numbers. The 'Bingo' craze owes something to the fascination of a pattern of squares with numbers in, gradually getting filled up. We believe that a desire to be able to use and understand numbers is built into mankind, just as is the wish to communicate by speech. It is surely wrong that this initial fascination is so frequently turned to distaste by the time that maturity has been reached.

Figs. 1.1 and 1.2 show two fascinating patterns of numbers. The first is the familiar Lo Shu of the Chinese. This is a number square, sometimes called a 'magic' square. The totals of rows, columns and diagonals have a relation to each other which it is not difficult to perceive. (See Answer 1 at the back of the book.) This is not of purely abstract interest, for it is similar to the 'Latin' square that the statistician will derive and use to distribute seed or seedlings between plots in statistical experiments. This is so that he can minimise the effect on his growth curves of small variations in soil or treatment between one plot and another.

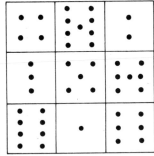

Fig. 1.1

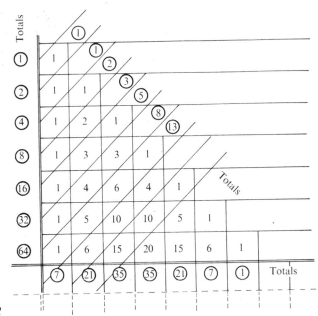

Fig. 1.2

The second pattern is incomplete, for there can be as many rows as are desired. The circled figures have been obtained by totalling horizontally, vertically and diagonally. There are interesting relations between the numbers in the various columns and in the numbers forming the top left to bottom right diagonals which have not been ruled in. The horizontal totals add up to the powers of the number two, i.e. 2, 2 × 2, 2 × 2 × 2 etc. The totals of vertical columns show an unexpected symmetry and all come to multiples of seven except for the final 1. The totals of the bottom left to top right diagonals form what are called Fibonacci numbers, which possess many strange properties and an interesting affinity with many natural objects. The table itself has applications in advanced algebra and in the theory of probability.

'numbers which possess many strange properties'

Both of these examples show how 'pure' mathematics and 'applied' mathematics link together and how the abstract fascination of numbers is matched by the utility of the subject. Try copying Fig. 1.2 without the totals. Add another horizontal row, it starts 1, 7, 21, . . . Then add up again but, before doing so, try to predict the totals of the rows, columns and diagonals. (This is quite testing! Do not worry if you cannot, but look at Answer 2 at the back of the book for help.)

Were you bad at mathematics at school?

Traditionally, mathematics is a subject at which it has been no disgrace to have been bad at school. Indeed we have heard this boasted of by those who should know better. The subject has sometimes been held to require special and rare aptitude which only a very few possess. It is certainly true that the genius occurs rarely, but this is true of the genius in all fields. It is certainly not true that only the specially gifted can appreciate or make use of the subject, as can quickly be seen by a visit to almost any classroom in which the ideas of the new mathematics are being investigated.

The key to this may lie in the value which society places on the subject. Until recent years it is true to say that it was a low value. Lip-service was paid to the necessity for mathematics as a 'qualification'; once this qualification had been gained then few were expected to make much use of the subject again, apart from specialists and, of course, the teachers. This had profound implications for the way the subject was tackled. Since it was the qualification that mattered, it could be held to be entirely logical to teach pupils principally how to pass examinations. It could seem to be of no particular interest or concern whether pupils liked the subject or really understood what it was about. Again it was felt by many that it did not particularly matter what syllabus was being taught provided it included simple arithmetic. Once a child had shown competence in learning the work contained in the chosen syllabus this was enough and it could all be forgotten.

It is scarcely possible to hold today that mathematics is irrelevant. A glance at the Sunday newspapers will show the enormous demand for the specialist mathema-

'A glance at the Sunday newspapers'

tician in many different industries. Nor is the specialist mathematician necessarily a graduate. Boys and girls with A-levels or good O-levels are required for programming computers or are being urged to try operational research or critical path analysis. Candidates for further education in languages, geography, economics and many other subjects are being urged to keep on their mathematics at school to the highest level that they can reach.

The days are past when the man in the street could ignore the subject. The small business man or shopkeeper may still prosper by *ad hoc* methods and make up his mind with little help from mathematics, but, for good or ill, this is the day of the large firm, the huge administrative district and the chain store. The enormous increase in complexity of industrial and commercial life requires properly calculated stock control, investment policy, market surveys and so on. The details are properly the province of the mathematician or statistician but the problems have to be posed and the answers turned into action by non-mathematicians. Decisions must be taken by directors, managers, foremen, and they too must understand something of the theory on which they are acting.

Electronic computers are being adopted and developed to help solve these very complex problems but computers are only as good as the minds that program and

'computers are only as good as the minds that program and design them'

design them, and to use them effectively it is equally necessary to have a feeling for mathematics.

A feeling for mathematics is an attitude of mind

Mathematical education today is judged almost as much by its success in producing a certain attitude of mind as by the body of understanding, skills and techniques it gives. It will have gone badly astray if the subject is looked upon as a miscellaneous collection of unconnected examples, each with its own little trick for getting the right answer. The attitude arises from the feeling that it a reasonable and logical study, worthy of pursuing for its own sake but offering much to the practical man.

The foundations of mathematics are its axioms. It is axiomatic that if two numbers are each equal to a third number, they are equal to one another. It is also axiomatic that through two given points you can draw one and only one straight line. Both of these propositions accord with common sense; no one will expect them to be 'proved'; such statements form the raw material of mathematics. Starting from axioms, the mathematician proceeds by the rules of logic, but he also builds up a most effective and labour-saving system of notation. This notation is the language of mathematics. Given this right attitude, the child will understand something of the why and wherefore of these axioms, they will be acceptable, being sensible and consistent. It will be natural to seek for a logical way of tackling situations and problems and, in many cases, it will seem natural to turn problems into mathematical language. This language comprises drawings and graphs as well as symbols, and it will seem reasonable that arguments should be made and cases put in this form. Anyone who says, 'My mind is made up, don't confuse me with the facts', or 'Of course, statistics can prove anything', has a totally wrong attitude of mind towards mathematics!

'Mathematics as a language'

Mathematics as a language

In many ways it makes sense to think of mathematics as a language with a vocabulary and a grammar. Before the development of language there could have been no, or very little, real thought. It is impossible to think or communicate mathematically without having a good grasp of mathematical language. It is highly condensed and for this reason much of what is written may look baffling at first sight. Indeed, with parts of mathematics which are new it is often the need to absorb the new language or notation which forms the biggest initial barrier to understanding.

Let us get down to examples. Here are three mathematical sentences:

1. $5 \times 4 + 2 \times 3 = 26$;
2. $a(b + c) = ab + ac$;
3. $(uv)' = uv' + u'v$.

You will be able to read and understand the first sentence. You may well recognise and understand the second sentence too. This expresses the rule for clearing brackets in the algebra of numbers. Unless you have learnt some calculus, the meaning of the third sentence, however, will remain obscure, and although superficially it may appear to be rather like the other two, in fact it is very different from them and you would need to study a good deal more before finding the key to unlock its meaning.

As well as being a highly condensed language, mathematics is also a very precise one. Humpty Dumpty said, 'When I use a word it means just what I choose it to mean, neither more nor less.' Here speaks the mathematician C.L. Dodgson alias Lewis Carroll. Professor Higgins in *My Fair Lady* puts the same thing in a slightly different and possibly unconscious way when he says, in the course of his song, 'Why can't the English teach their children how to speak?' that the French don't mind what you do so long as you pronounce it properly. Both Humpty Dumpty and Professor Higgins are making the point that there is no use starting an argument until you know what you are talking about. Once you have made up your mind what you are talking about, then, and then only, do the processes of logic and analysis start, and from then onwards everything is on a proper basis.

It is when business firms come to analyse their sales policy or when a manufacturing firm seeks to find the proper site for its new factory that it becomes vital to analyse objectives, define conditions and say what is meant by a 'best solution'. The language of mathematics can be translated into the languages of business or technology.

'It is when business firms . . . analyse their sales policy'

Mathematics is creative

The most exciting part of all mathematics is to solve a problem of one's own formulation by a method invented for the purpose. If mathematics seems to consist entire of solving problems in textbooks, the answers to which are in the back, it ceases to stimulate students in the way that it can and should. From this point of view it matters little whether the problem is a trivial one about noughts and crosses or a deep one about nuclear science. The feeling of mastery and success is much the same. More will be said later about bringing creative activity into a child's education at an early stage, and the possibility of inventing one's own problems.

'it matters little whether the problem . . . is about noughts and crosses or . . nuclear science.'

'Modern' mathematics

This is the era of the catch phrase and there has been a general desire to find one to describe the new syllabuses in mathematics and the new approach. The most commonly used terms are 'modern' mathematics, 'new' mathematics, sometimes 'contemporary' mathematics. All these are misleading because a great part of the 'new' mathematics now being taught is in fact at least a hundred years old. What is new is that it is now being taught in schools. This is the result of the fact that mathematics is a quickly changing subject. It has been stated that there has been more mathematics discovered since 1950 than in the whole of the previous history of man. The mere weight of all this new knowledge and discovery has tended to push subjects downwards from the universities into the secondary schools and from the secondary schools into the primary schools. One of the revelations has been the extent to which subjects previously thought of as being highbrow and deep can be mastered at a much earlier stage. Children in secondary school today can tackle problems in arithmetic which would have been challenging for a teacher of mathematics in classical times.

A simple calculation once difficult

As an example, let us consider long multiplication.

$$
\begin{array}{r}
25 \\
\times \ 13 \\
\hline
250 \\
75 \\
\hline
325
\end{array}
$$

The presence of the zero in the first line of working shows how the decimal notation permits 'automatic' multiplication by 10. The rule that 25×1 is offset one place to the left makes the answer into 250, that is, 25×10.

Compare this with the same calculation done in Roman numerals:

$$
\begin{array}{r}
XXV \\
\times \quad XIII \\
\hline
CCL \\
XXV \\
XXV \\
XXV \\
\hline
CCLXXXXXXVVV
\end{array}
$$

= CCLXXXXXXXV	(since VV = X)
= CCLLXXV	(XXXX = L)
= CCCXXV	(LL = C)

The big differences are that there are no 'columns' to help organise the work, and that it is necessary to convert the answer into acceptable form. Remember that VV stands for V + V in Roman numerals so that VV is X. The difference in division is even greater; just try it!

There is a still greater contrast between ancient and modern skills in algebra. Problems of the 'Find the number' type are now easily solved by a child who has learned the algebra of numbers. Before the invention of modern symbolism, problems of this type taxed the best mathematicians of the day.

Specialists or general practitioners?

The need for specialists in mathematics of varying degrees of excellence has, of course, increased greatly as a result of the increasing dependence of government, industry and research on mathematical skills, and, in particular, on the electronic computer.

It is now becoming increasingly apparent that many studies, previously thought of as unrelated to mathematics, in fact can make great use of the subject. Many university departments of chemistry are requiring advanced mathematics as a condition of entry. Indeed, at Oxford the only examination taken in the first year of a chemistry course is one in mathematics. Many schools of biology, economics, geography, social science, etc., need statistical techniques, and biochemistry needs probability theory. Crystallographers and physicists are finding the need for some of the deeper ideas of abstract algebra, and an application of the theory of 'groups' has recently led to the discovery of a new and unsuspected elementary particle of matter. Computers are now being used for weather forecasting, automatic translation from a foreign language, the compilation of school timetables, and many other things. These all involve many professional people who are not in any sense specialist mathematicians but who need to understand the language and some of the techniques of the subject.

This all adds up, we think, to a formidable case for a general mathematical education which makes sense in relation to present-day needs. Mathematics has importance both in the personal and in the social education of the individual. Because of this there has been an enormous increase in the demand for general practitioners of mathematics. Much of the driving force for the new mathematics project has been a desire to improve the mathematical education of the ordinary child rather than of the future mathematical specialist.

There are signs that this is being recognised in all quarters. It was no accident that the first series in the British Broadcasting Corporation's educational television broadcasts on BBC 2 was called 'Mathematics '64', and dealt with new materials both for the teacher and for the parent. Voices in the university world had been heard recently calling for compulsory mathematics up to school leaving age, whethe this be 16 or at the end of A-level courses. Articles on mathematics have appeared in national and local newspapers, and even in some of the glossy magazines. Advertisers no longer avoid the statistical approach to the selling of their wares (although this is not always handled very well!).

Mathematical education

The aim of mathematical education can therefore be stated very easily. It is that more people shall know more mathematics, and shall know it more thoroughly. This demands better teaching, certainly, but also better learning. This is easy to say, less easy to achieve.

In the following chapters we shall try to show you something of how it is being done and we start by describing and discussing some of the new things being done in schools. While we shall concentrate on the new, it must not be forgotten that basic arithmetic, simple geometric ideas and the concepts of algebraic language remain the backbone of mathematics — the same for all time, comfortably familiar!

We are also going to involve you in the mathematics, as well as talk about it, and to answer the question 'Why?' at the same time as the question 'How?' The cards are therefore on the table; this is an educational book, as well as a book about education. You have been warned!

2 Project work and individual learning

Changes in the content of their mathematical syllabuses have been accompanied in many schools by efforts to end the separation of the more able from the less able children, so far as is possible. An exception is made of the very slow learner, who receives special help, and it is a matter of much debate whether there is also need to give special treatment to the extreme upper end of the ability range.

Where children of widely differing ability are being taught together, the 'chalk-and-talk' lesson probably familiar to you is inappropriate; it may bore some children and baffle others.

Projects

One way of dealing with a mixed ability class is to teach through projects. This means that each child is asked to select a topic that interests him (or her), for example, Tibet, or travel, or the development of the number system, or bus services, or Ancient Britons. He then gathers information from whatever sources he can and writes up his findings in an attractive manner. The sources from which the information can be gleaned are legion. Nowadays they are generally called resources, and are kept in a resources centre, often the room or building previously known as the library.

The main resource is still the book. Book production has improved tremendously with the widespread use of colour, and with more and more attention being paid to design. Today it is much more fun for a child to look things up than ever before, but it is also much more complicated because of the increase in size of many school libraries and in the multifarious nature of the books available. Potted knowledge can, of course, be found from encyclopaedias and from books which review a whole area, like books on travel or transportation. Some of these consist mostly of pictures with a minimum of simply-written text, while at the other end of the spectrum there are scholarly tomes with much economic data, maps, analyses and so on. A worthwhile project will demand that the child reads a number of books, some only making passing reference to the topic that he is trying to write about. Much information can also be found in magazines, brochures, newspapers and the like. The problem of cataloguing these, together with books, for maximum ease of access is very difficult. Again, some firms publish collections of material, such as facsimiles of original documents, old timetables or tickets, copies of articles from publications, handbills, etc. To add to the list of printed material, there are maps, statistical tables, chronological summaries and so on.

Also in the resources centre will be found much non-printed material. There may be gramophone records or tapes of music, drama, natural sounds which may be relevant, or perhaps a tape of a talk on the radio. At some schools teachers can make video tapes of television programmes to play back later. There are almost always slides to be looked at in a slide projector or film loops which can be played back on a special machine.

All this is second-hand material. For a project there may also be a certain amount of basic research needed, for instance, a survey of actual local traffic near the school, or a visit to a factory or a dig at the site of an archaeological settlement.

Perhaps enough has been said to convince you that it is not exaggerating to call this research, and that the project method of learning can be a most demanding one. The child has to find his way through all the material referred to, disentangle the relevant from the irrelevant, reconcile possibly differing views or statements, and then find the best illustrations or make the best model, in doing which he is learning a skill of great significance for his mental development.

'it is not exaggerating to call this research'

Obviously the depth to which this treatment is taken varies very widely. This is precisely why the project method goes so well with the abandonment of ability grouping. Indeed, it is hard to know which, here, is the chicken and which is the egg, for the decision to teach through projects can make such grouping largely irrelevant, since no two children in the group will be doing the same thing in the same way.

In addition to learning how to select, judge and present, there is great scope in the project method for making use of basic skills. The text has to be written. Good grammar, spelling and selection of appropriately vivid words are essential. Visual materials and drawings have to be prepared, calling on craft and art skills. Statistics have to be assembled and criticised, and calculations have to be made, sometimes with the aid of machines, calling for mathematical ability and understanding. It is possible of course for these skills to be present only at a low level. A good teacher will find ways to stretch what there is, and to upgrade skills as experience is gained. There seems little doubt that a project which fires a child's imagination does serve as a good motivation for his or her development in very many directions.

It must be understood that the project method can result in a changed role for the teacher. From being the figure of authority who dispenses knowledge, requires that it is learnt and tests to see that it has been, the teacher can become more of a guide, a partner with the child in the development of the project.

This is perhaps the point at which to observe that this method makes much greater demands on the teacher than the more conventional teaching method. Instead of having merely to know his own subject, the teacher has also to have a thorough understanding of the working of the resources centre which serves the school, and to have some acquaintance with each of the basic skills mentioned above. He has also to do his own private reading on the topics chosen by the thirty or so children in his group, so as to be able to keep a careful check on the progress of each, and offer advice where necessary. The average child needs constant encouragement, and it is not satisfactory for a project merely to be looked at when finished. The teacher must be brought into it at each stage of building up the project. He must keep a critical eye on the preparatory reading that the child does, he must see that the subject-matter is properly organised, and, if necessary, help the child to work it out. All this makes heavy demands on his time, patience and skill as a teacher.

Self-paced learning

Another way of dealing with a mixed ability group is by self-paced or individualised learning. What would have been a conventional textbook is rewritten in small sections, usually confined to one idea, and printed on cards or in small topic booklets, or sometimes recorded on film loops or tapes. A certain amount of freedom to investigate or experiment may be built in, but in the main the work is structured. This means that the work is set out in a definite sequence, with later work building on earlier, and, in general, being harder.

With this system, each child is face to face with an author. Learning is dependent on the writer's skill at explaining what he means and what he wants done. Mainly he will rely on words, but he may also use symbols and often pictures or diagrams. For the poor reader there may be tapes so that he can listen to what is required. There is a real problem in gauging the reading skill of the children he is writing for, and using words and ideas which are within their compass as well as being exciting and interesting to them. Most modern textbooks – including the School Mathematics Project (SMP) series – have been written with this in mind.

There remains the problem of dealing with the various levels of ability that will be found in every group. One way is to have a series of booklets which deal with the main topics of a course in a fairly simple way: we may call this the 'A' series. For instance, for the eleven-year-old there may be booklets entitled: 'Addition and Subtraction of Fractions', 'Binary Arithmetic', 'Set Notation', 'Symmetry and Reflection', and so on. Each one contains straightforward instructional material, together with worked problems and short self-tests, answers being immediately given. There are further questions whose answers may be given 'on a later page', or not at all. There may also be open-ended questions – that is, questions of the following type:

'Draw a letter L and its reflection in a line. Reflect the reflection in a parallel line. What do you notice? Draw some more parallel lines and reflect the letter L over and over again. What happens? See what happens if the lines are (*a*) close together, (*b*) far apart. Draw lines in different directions and see what you can find. Would it matter if the lines were not straight? Find out some more about repeated reflection.'

In a question like this the child can take the matter a lot further or a little; he may develop his own lines of thought and follow ways of thought that the author has not conceived.

When a child has worked his way through a booklet, he will take a test-card. Here he finds a series of questions designed to test his understanding of the 'A' series booklet. He works his way through, then gets an answer sheet and marks his own work. If he has done well enough (and the teacher will say what 'well enough' means, perhaps, 75% correct) he may move on to the next booklet in series 'A'.

If he gets stuck while he is working through a booklet, or if he fails to score enough marks in his test, he takes a 'B' series booklet. This contains remedial work. For instance, if he had the 'A' series booklet entitled 'Adding and Subtracting Fractions', the 'B' series booklet would be entitled 'Fractions Revised' or 'Fractions made Easy'. It would go back to basic definitions, make diagrams to illustrate fractions, show how fractions combine, illustrate why $\frac{1}{2} = \frac{3}{6}$, and so on. It would be written at a simpler level.

When he has finished working through the 'B' booklet, he tries the 'A' booklet again. Probably he will go through rather faster this time, and he then tries the same test again and, it is to be hoped, passes on this occasion.

Some children will be forging ahead at a great rate and demanding a new 'A' series booklet almost every day. Usually the teacher will not want too much divergence between the broad stages reached by the slowest and by the fastest in a group, and there is therefore also a series 'C'. These are sometimes called enrichment booklets, and enable the quickest to keep their interest and increase their skill, being based usually on similar material but going into greater depth.

For example, there might be a booklet on the fraction work of the Ancient Egyptians. These people worked with fractions with a unit numerator only. These are fractions with 'ones' on top, of the form $\frac{1}{5}$, $\frac{1}{31}$, $\frac{1}{100}$, and so on. There was just one exception, the fraction $\frac{2}{3}$, which curiously they did use; they used no fractions like $\frac{3}{4}$ or $\frac{7}{8}$. They managed to build pyramids which were aligned to an extraordinary degree of accuracy without, for instance, being able to say, 'I need this length to be seven-eighths of that length.' What they would have to say would be, 'I want this length to be a half plus a quarter plus an eighth of that one.' A quick check will convince you that $\frac{1}{2} + \frac{1}{4} + \frac{1}{8} = \frac{7}{8}$.

The teacher must know each child's progress and the child himself will keep his own record. This is done on a card which will show a pattern. It may be a straight progression through the 'A' series booklets, or there might be a pattern consisting of an 'A' booklet followed by a remedial 'B' booklet, and so on. Another possibility may be a set of 'A's punctuated with the enrichment booklet 'C's. From the record card it is quick and easy to gauge a child's ability and such a record card almost constitutes a school progress report.

Behind these new methods of teaching and learning is the argument that academic ability is present in everyone — but while a few achieve results in a short

'academic ability is present in everyone'

time, most may achieve similar results given longer, in some cases a great deal longer. Hence making careful assessment of a child's rate of working becomes extremely important and so does providing the right experience and material at the right time.

Obviously this idea is a difficult one to demonstrate. It is certainly far from universally true that, given time, all children can achieve the same goal. Motivation, for instance, is of paramount importance and may be lacking. Where it is, neither ability, nor length of study will produce academic success. Again there are some marked differences in types of intelligence. For some, number presents no difficulties whatever, for others, number seems to make no sense but, perhaps, there is good visual perception, or skill with words.

With project work and individual learning, new methods of organising timetables, classrooms and teachers' time assume great importance. Traditional views of the appropriate levels of examination may need reconsidering too. If the 'time required' theory is even approximately true, it may be inappropriate to expect examinations to be regularly taken at any particular age. It would be necessary to allow higher education to start at any age, perhaps well into adulthood. A step in this direction can already be seen in Britain with the advent of the Open University. This uses public broadcasts and a nation-wide net of part-time tutors to make degree courses available to anyone with the time and inclination to study.

In England and Wales, 'public examinations' are closely related to age. These are the General Certificate of Education (GCE) at Advanced (A-) or Ordinary (O-) level and the Certificate of Secondary Education (CSE). A-level is linked loosely to age 18, but some O-level and CSE boards require a signed certificate that a candidate · has reached the age of 16, or, if slightly below, has taken the necessary courses and is ready for the ordeal. It is common to talk of the 'three A-level' student or the 'one A-level' student as though the former were a different type of being from the latter, being intellectually capable of keeping three academic subjects (the normal load) on the go at the same time, as opposed to the intellectually less able student who can only cope with one. It may be that this year's 'one A-leveller' will be next year's 'three A-leveller'! However it may also be significant that reducing the study load by giving up one or two of the three A-level subjects is far from guaranteeing better performance in what is left.

We remarked above that more is demanded of the teacher when projects have to be supervised or a class is working at the whole range of self-paced work. More too is demanded of the child and this brings its problems. As there is scope for more and better work, so there is scope for more idleness and even mischief. It is no use pretending that every child is fascinated by every project that he chooses or that every child will have his imagination fired by every card or pamphlet he picks up. There are plenty of difficulties also in planning revision of material and in ensuring regular review of concepts gained and facts and skills learned so that they may not be lost again. Much work is being done to improve the new methods and to provide new teaching material. But all is not yet as good as it might be.

It is only fair to add too that, particularly in English and mathematics (the two basic subjects) voices have been raised against new methods as well as in their favour. As a new publication from the office of the School Mathematics Project makes clear, it is possible to get the balance wrong, certainly in mathematics. It is not enough merely to stimulate pupils and to make them excited with their work. They must leave school able to perform the tasks society expects of them so that they may perform their jobs competently, whether they be computer operators or bus conductors.

Against this must be set our firmly held view that the twin dangers of distaste and boredom, that demonstrably made mathematics the most hated and feared of subjects for many people in the past, may be infinitely worse than any small diminution of manipulative skill. Not that we think this does not matter. We back modern mathematics and modern methods because they make for more and better mathematics, not for less and worse!

In the next chapter we describe another key idea in the new mathematics teaching — the 'discovery' method. This particularly helped the rejuvenation of much of the work in the primary school, to which we now turn.

3 Changes in the primary school

Teaching or learning?

We make no excuse for starting this chapter on the primary school with the changes in teaching method and leaving the content of new mathematics courses until later. Many teachers would subscribe to the view that, within reason, the syllabus is less important than the way in which it is taught, and that simply producing a new school syllabus for primary and secondary schools is merely tinkering with the problem. What is wanted, they would say, is a new philosophy. Nor would we apologise, secondary school teachers as we are, for talking about primary schools. It is of supreme importance to make sure that the foundations of the subject are properly laid, and our primary colleagues will surely forgive us for venturing into their territory!

The task of the teacher

The teacher in the classroom is the basis of the whole learning process. We would suggest that his (or her — but we will not repeat this each time) proper task can perhaps be summarised under three headings:

(a) to motivate;
(b) to set the scene against which the children may learn, both in a structured and in an unstructured way;
(c) to present information.

It is interesting to ask teachers to put (a), (b) and (c) in order of importance. Although they will often (and rightly, in our opinion) place (a) and (b) before (c), traditional teachers have often spent a large proportion of each period on the aspect of education which they consider least important, that is to say, they feel it their duty to present a great deal of information. It is obvious that presentation can be made in many ways and on many levels. It is not difficult to present facts to many children at the same time and certainly it is this way of teaching which is simplest for the teacher.

Concepts

It has long been realised that there is a considerable difference between learning a fact such as 'the Battle of Hastings was fought in 1066' and the understanding of a concept, such as the concept of a number. Once the date of the Battle of Hastings has been given, it is accepted as a fact, and is either remembered or forgotten. It will not be possible to answer the question, 'What is a number?' in such a way. It is of no use to reply, 'Well, numbers are 1, 2, 3, 4, 5, 6 . . . , that's what numbers are. Now you go off and learn them up to 100.' There is not much use, either, in chanting number relations such as '3 and 1 make 4, 3 and 2 make 5, 3 and 3 make 6, 3 and 4 make 7', if the meanings of the words 'and' and 'make' in this context are not appreciated.

How would one set about helping a child to understand the number 2? Perhaps one would start by holding up two bricks and saying, 'Look, 2'. This might be followed by holding up two fingers and saying, 'Look, 2'. Why is it necessary to hold up the fingers as well as the bricks? Because the number 2 does not depend on the bricks or on the fingers; it is an abstraction from them. It is a thing that two bricks and two fingers have in common. In primitive tribes, and some not so primitive (for example, the Greeks!), it is often the case that there will be a special word for two fingers and a different word for two bricks. They have not made the mental abstraction of realising that '2-ness' is something they have in common. When the child has seen many examples of pairs of objects associated with this word, he may still think that '2-ness' is a property of concrete objects and not be ready for the '2-ness' of two ideas or two seconds. There is no point in going on to the techniques

' *"2-ness" is a property of concrete objects*'

of addition or subtraction until this concept has formed in the child's mind. Of course, once he has formed the concept 2, the concept of 3, 4 and 5 will be that much the quicker. The next step is to understand that 2, 3, 4 and 5 also have something in common: they are all numbers, counting numbers.

It is perhaps significant that the word 'teach' has not occurred in what we have said so far. Teaching is a word for telling or showing: you tell someone that Nelson's Column is in Trafalgar Square. Facts can be taught but it is not possible to 'teach' proper understanding of number.

Notation

Closely connected with understanding the concept of number is finding a way to represent numbers. The number 2 is independent of the objects, be they bricks, fingers, buns, ideas or ice-creams. It is also independent of the symbol used to express it. We write the number '2', the Arabs, 'Υ', the Romans wrote it 'II'. The actual symbols used to represent numbers are usually called numerals. There is an essential difference between the Roman system of numerals and ours and the Arabs' Whole numbers plainly go on forever; but we use only ten numerals for them all; these are the symbols 0, 1, 2, 3, 4, 5, 6, 7, 8, 9. In the Roman system, however, the numerals themselves also go on and on, for although the numerals I, II, III, IV, etc., can be used over and over again, extra, different letters are used to express 100 (C), 1000 (M or CIƆ), 10,000 (CCIƆƆ), 100,000 (CCCIƆƆƆ) and so on, so that there is no limit to the number of symbols that could be needed.

Mathematical concepts are not concerned only with number: it is also necessary to understand ideas such as length, volume, angle, equality and inequality, relations between sets, etc. As each new concept is met, new notation is often needed.

The 'discovery' method

Many teachers would agree that an ideal situation for learning is to give the children suitable materials, for example, pieces of apparatus, experiments to perform, numerical operations to carry out, etc., and by carefully planned questions lead them to discover mathematical relationships for themselves and so build up their own concepts. This is obviously quite difficult for those who create the questions and the teachers who use the 'discovery method'. They must be certain that each child does form the concept for which the material was prepared and that progress is made at the best pace each child is capable of.

This poses something of a dilemma for us (the authors of this book)! Since you (the reader) should obviously be dealt with in the same way, our objective should be to lead you gradually through the notions of mathematics which you will discover for yourselves as time goes by. Within the limits of this small book, however, this is scarcely practicable, and we lack the essential classroom contact, so you must bear with us if we treat you in a more old-fashioned and direct manner.

Learning can take place most readily when the classroom is generously provided with practical material, carefully selected and matched to interests and activities so

Fig. 3.1

that it bears on the topic in hand. As an example, let us consider the idea of volume. For this it will be necessary to provide the class with measuring jugs, liquid – and a plentiful supply of material for mopping up (Fig. 3.1). Questions may be printed on cards or in booklets or may spring from the project in hand, perhaps guided by the teacher. That volume is quite a complex idea may be revealed by asking older children, or even adults, to define what exactly they mean by the word. The volume of a biscuit-tin may perhaps be described fairly easily, but what about the volume of a walnut or a human being? And there is more to it than that. It came as something of a shock to teachers when Piaget proved that it is not obvious to a young child that volume is conserved. This means that the quantity of water in the water-jug is not

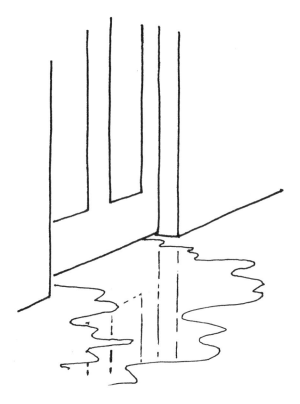

'It came as something of a shock to teachers'

changed by being poured into a number of glasses. Fig. 3.2 shows a short fat
measuring jar and a long thin one. A child who has not appreciated this concept is

Fig. 3.2

very likely to say that a pint of water poured into the latter is 'more' than another
pint of water poured into the former. He is confusing the depth with the volume.
Any attempt to measure the volume inside something irregularly shaped, such as the

walnut shell, will plainly consist of pouring the contents into a graduated cylinder, and is doomed to failure if the child does not realise that the volume does not change.

Understanding what volume is, is closely related to understanding how volume is measured. It is easier to measure and illustrate area, so let us consider this concept in more detail. Following early experience of large and small surfaces, it is a common development in the primary school to invite children to bring leaves to school. They will bring them from a variety of trees, shrubs or plants. The problem is to try to decide who has brought the largest leaf. Discussion starts. (You will notice

'try to decide who has . . . the largest leaf'

that we take it for granted that this will happen!) Many ill-conceived ideas will be forthcoming. Do you agree that this is a good way to start? Before reading on, think about the problem of comparing areas.

In the first place it may seem that those who have brought the long pointed leaves, like those of the laurel, will win, but doubts may creep in when children arrive with broad leaves from the water-lily, or finger-shaped ones from the lupin (Fig. 3.3). Tracing the leaves onto a piece of paper makes the problem look slightly

Fig. 3.3

more formal, but there will not normally be agreement on how to tell which is the largest of the leaves shown.

An idea often proposed is to measure the distance round the outside, that is to say, to measure the perimeter. Thinking about the lupin leaf usually convinces children that area and perimeter are not directly connected. Tracing onto paper with a pattern of dots or squares or triangles suggests answers to the problem (Fig. 3.4).

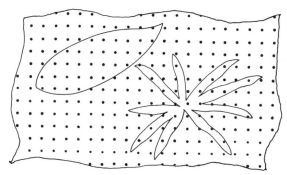

Fig. 3.4

Count the number of squares or dots or triangles and see which one contains the most.

It is important to move the tracing to new positions so that the approximate constancy of the answers can be seen. Objections will be made about the fractions of units towards the edge, and suggestions may be made, perhaps, for counting fractions. A more fruitful idea, though, is to take a new piece of paper with dots closer together or covered with smaller squares or triangles. Any formal definition of area depends on some concept of limit, that is to say, one supposes the size of the pattern or grid to be made smaller and smaller, so that the fit gets better and better. This is a sophisticated idea, however, and it is not appropriate to push this to the point of giving a formal definition. Traditionally, the only work done on area in primary schools has been the learning of various formulas, such as 'length times breadth' for a rectangle or 'a half base times height' for a triangle. This was surely putting the cart before the horse, for until the meaning of an area is understood there can be little value in learning formulas.

The classroom in which mathematics is being learned in this new way will rarely look very tidy (Fig. 3.5) and almost certainly be rather noisy. Much of what the children are doing may look as if they are playing. This is not accidental; it is now realised that play is an essential part of the learning process. But the apparatus and the games played should not be random, but be carefully thought out. All young animals play and, in the course of play, rehearse real situations which will confront them later on. Play is the child's way of making trial of the world around him, of gaining experience, and of forming a private picture of the universe. Whereas geography and biology are all around us (it is impossible to go outside the house

Fig. 3.5

into the street or garden without experiencing something of them), mathematics is abstract: experience must be arranged.

We emphasise again that this play has to be 'structured'. At any given time, the apparatus to hand should be limited to what is required for understanding the topic being considered, for example, volume, or area, or large numbers. In too mixed an environment a child may get confused. Again, the right questions need to be asked, if not by the child, then by the teacher or writer. The teacher has to step in at the right time to make sure that the problem is properly understood, and that the child does not waste a large amount of time by going off on the wrong track. In the case of the area of the leaves, for example, he must be ready to show that 'putting a string round the edge' will not do. The perimeter is not related to the area. He might do this by showing what happens to a beech leaf whose perimeter can easily be increased by making a cut in one of the sides, but whose area, of course, is unchanged by this process. You see the necessity for understanding conservation before dealing with questions of size and the comparison! Nor will the child, unguided, always realise when he has reached the answer. A child's activity is not always purposeful and is usually not well organised. If the right questions and summarisation are not forthcoming, the child's activity may all be wasted. The teacher must also be alert for signs of exasperation or boredom, for these may denote unreadiness, a problem beyond the child's powers to solve or an idea too hard for it to understand. Then he must be ready to suggest new ideas or to change the subject

'signs of exasperation or boredom'

Structural materials for understanding addition and subtraction

Structural materials are very useful in that they can isolate certain aspects of mathematics (for instance, the four rules) and help children to gain the experience from which understanding comes. A well-known example, found in almost all primary schools, is a set of graduated wooden rods, several varieties of which are available commercially. They normally have square cross-sections and the smallest has the same length as the side of the square, that is to say, it is a cube, and this is called the unit. Other rods come in lengths which are multiples of the length of the side of the square, and some examples are shown in Fig. 3.6. With some materials, the numbers of units are shown by markings on the rods, as in the diagram. In others there are no markings, but different-length rods may be in distinctive colours.

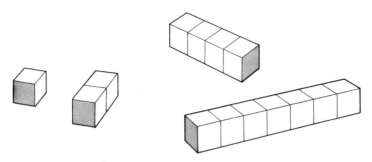

Fig. 3.6

The rods are usually called 'ones', 'twos', 'threes', etc. Fig. 3.7 shows how a 'three' rod and a 'four' rod are put together and matched with a 'seven' rod. This is a practical demonstration of the fact that 3 + 4 = 7. From this, and a large number of other similar experiments, the child builds up his idea of addition. He will get a

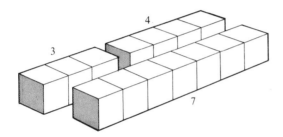

Fig. 3.7

different view of addition when he puts together groups of three and four different objects, for example, model cars. The essential difference between the two is that with separate objects like model cars it is necessary to count them each time to make sure. When the rods are used and the child is familiar with their use, the counting is not needed, for he can judge which is the 'four' and which is the 'three'. The concept of addition is consolidated by adding in different ways.

This difference may perhaps be seen even better in the context of subtraction. There are two types of subtraction, and the model cars illustrate the first type (Fig. 3.8). With model cars, 4 − 3 is visualised as removing three cars from the group of four, leaving, of course, one.

The second type of subtraction is best illustrated with rods, 4 − 3 = 1. Using the rods, Fig. 3.9 shows that what is wanted is a rod to put together with the 'three' to make up the 'four' rod. After some experiment, the child will see that it is the 'one' rod that is required to fill in the gap.

Of course, it is important that the child should see each of these aspects of sub-

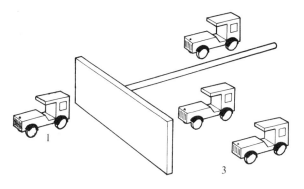

Fig. 3.8

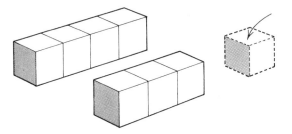

Fig. 3.9

traction as abstract, not something to do with rods or model cars. You might like to spend a moment thinking how rods could be used to demonstrate the 'removing' nature of subtraction. It is, of course, a sawing-off process. What will happen to the 'four' rod if a section corresponding to the 'three' rod is sawn off? The answer, once again, is that a 'one' rod is left. It is interesting to note that it is the second, or building-up, type of subtraction that is normally used for giving change. When you offer a tenpenny piece for something which costs three pence, the shop-assistant will generally count on from three pence and say, three pence, five pence, ten pence, handing you the appropriate coin each time. At the end of this you will have received a twopenny and a fivepenny piece totalling seven pence, representing the answer to the subtraction, ten take away three. It is interesting to note that the shop-assistant never has to arrive at 'the answer' to this particular problem. Curiously enough the traditional method of subtraction that the child learns at school will not assist the grown-up shop-assistant to give change in this way. Indeed, he may not always realise that he is doing a subtraction when giving change! This is one of the many places where a more rational approach to the foundations of the subject should build up an understanding which will carry over into everyday adult life.

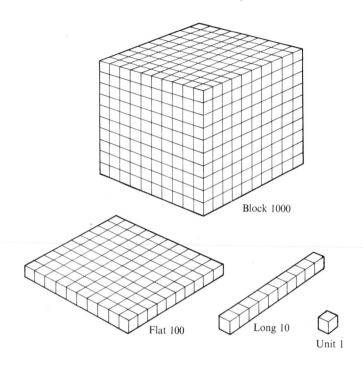

Block 1000

Flat 100

Long 10

Unit 1

Fig. 3.10

Place value

There are two main types of structural apparatus for helping children with this, one is shown in Fig. 3.10 and is an extension of the rod system to 'flats' and 'blocks'. The other is a simple abacus. This may be of the type where beads move on parallel wires, or it may be a spike abacus of the type shown in Figs. 3.11 and 3.12. It consists of a base-board, on which is mounted a series of spikes of equal length. On each of these a certain number of discs may be threaded. This apparatus is intended to enable the child to build up a concept of place value, that is to say, he has to understand that the meaning of the '2' in the number 52 is different from its meaning in the number 25.

How does the spike abacus help to form this concept? Well, we usually start with the idea of keeping a tally. Suppose we wish to record the number of exercise books given out to a class of thirty. We may suppose that we have, say, 18 discs, and that the spikes have been adjusted in length so that each one holds exactly 9 discs and no more. As each exercise book is handed out, one of the discs is placed on a spike.

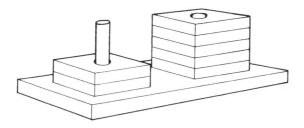

Fig. 3.11

Fig. 3.12

Once 18 books have been handed out, of course, the spikes are both full and the discs used up. What do we do now? The first suggestion will always be to get another spike, and more discs; but this is not necessary. Supposing that a disc on the left-hand spike means something different from a disc on the right-hand spike; what might it mean? An acquaintance with pounds and pence, weeks and days, metres and centimetres, may suggest that there are large units as well as small units. In class, it may or may not need to be suggested that perhaps a disc on the left-hand spike might be 'worth more' than a disc on the right-hand spike. Does it matter how many a disc is worth? In principle, no. Suppose that a disc on the left-hand spike is worth 3, then we have two different ways of exhibiting the fact that three exercise books had been issued. We could either place three discs on the right-hand spike or one disc on the left-hand spike. The number 18 will be represented by six discs on the left-hand side but this does not relate in any way to the numerals we use and we are stuck on 36 (nine 3s and nine 1s). Is there a better value? Often a child will suggest making the left-hand spike worth 18, to see what happens then. You answer, 'We can count 1, 2, 3, 4, 5, 6, 7, 8 or 9 exercise books by placing discs on the right-hand spike, and we can show 18 by placing a disc on the left-hand spike. What shall we do, however, about 10, 11, 12, etc.?' Further questions and answers of this type may enable the children to see that we want a disc on the left-hand spike to mean 10 units; thus a single disc on the left-hand spike carries on from the maximum number of nine discs on the right-hand spike. There is now only one way of showing the number 10: one disc on the left-hand spike. Fig. 3.11 shows the number 25. It has two discs on the left – two 10s – and five on the right – five units. We now see that it is possible to keep a tally up to 99, that is to say, nine 10s and nine units.

It is interesting to note that such forms of tallying did, in fact, represent the simplest forms of counting. For very many purposes it is not necessary to know the actual number of items that one is concerned with. A shepherd making sure that he has his sheep, simply wants to know that the number when he goes home at night is the same number that he took out in the morning. If he carves a notch in a stick for every sheep that he takes out, he has simply got to match the sheep that he takes back with the notches. It has long been a bit of useless information in the Tammadge family that sheep farmers in Surrey used to count as follows: onethrum, twothrum, zinneram, zanneram, cockerum, corum, tardiddle, wineball, zigtail, ten. What happened if a shepherd wanted to count more than ten sheep? It might have been at this point that he made a notch in the stick for ten and started again. Of course, the idea of matching objects with discs or notches in a stick represents the very basic idea of counting. The notion of the counting numbers has finally been understood when the child sees that real objects can be matched with the abstract things that we call numbers.

Bases other than 10

There is no need for the spikes to hold just nine discs each. It is possible to adjust

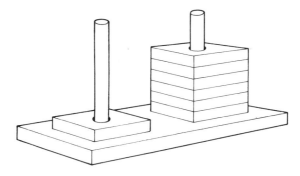

Fig. 3.13

the length of the spikes so that they hold three, or six. In the latter case, it is necessary for the left-hand spike to represent the number 7; then one disc on this spike and six on the unit spike, as shown in Fig. 3.13, represent the number 1 × 7 and 6, that is, 13. This is known as working in arithmetic to *base 7*. No one works arithmetic in base 7 from choice, but one disc on the '7' spike and six others can correspond to one week and six days. One week and six days added to two weeks and four days amounts to four weeks and three days. This is arithmetic in base 7! Working in many bases (multi-based arithmetic) strengthens the understanding of place value and highlights the value of base 10. Many teachers believe that this understanding should precede the learning of tables, and several different types of structural apparatus are designed to function in many bases.

Binary arithmetic

A very interesting arithmetic, and one that your children may well come home talking about, is that in which each spike holds only one disc; thus a disc on the left-hand spike has the value 2. Counting then proceeds as shown in Fig. 3.14. It is plainly necessary to introduce a third spike pretty quickly; in fact, as soon as the other two are filled, which occurs when we have counted as far as 1 × 2 + 1 = 3. Discs on the third spike, therefore, count 4. What will one count on the next spike? Counting on tells you that each disc on this will be worth 8. Perhaps you had noticed the following:

Values of discs on spikes in binary arithmetic

| 8 | 4 | 2 | 1 |

Values of discs on spikes in decimal arithmetic

| 1000 | 100 | 10 | 1 |

You have, we are sure, noticed that in binary you are dealing with powers of 2, whereas in decimal you deal with powers of 10.

Side view of abacus	Represented by	Decimal number
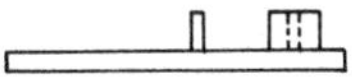	1	1
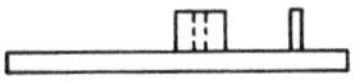	10	2
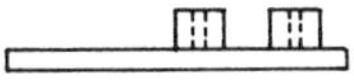	11	3
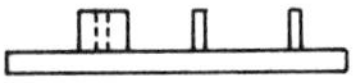	100	4
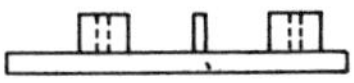	101	5
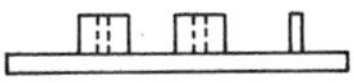	110	6
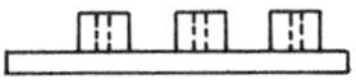	111	7

Fig. 3.14

It is not just a matter of counting in bases other than ten. As we have seen with weeks and days (base seven) we can calculate in them as well. Here is a calculation in binary arithmetic:

$$
\begin{array}{r}
^1\;\;1\;\;1 \\
+\;\;\;\;\;1\;\;0 \\
\hline
1\;\;0\;\;1
\end{array}
$$

You will recognise this as an addition. Look at it column by column from the right.

First $1 + 0 = 1$; that is easy.
Then $1 + 1 = ?$ you see that the 'put down' digit is 0 and that 1 has been carried (the small 1, top-left).
This is because $1 + 1 (= 2) = 10$, that is, one 2 and no units.

A longer addition sum goes like this:

$$
\begin{array}{r}
^1\;\;\;\;^11\;\;^11\;\;1 \\
+\;\;\;\;\;1\;\;0\;\;1 \\
\hline
1\;\;1\;\;0\;\;0
\end{array}
$$

In the third column we see $1 + 1 + 1 (= 3) = 11$, that is, one 2 and one unit.

In computing, binary digits are called 'bits' (*bi*nary digi*ts*). Binary counting is of importance in the design of computers, since electric circuits are most conveniently arranged to be in one of two states, on or off, high voltage or low voltage. Some people suggest that binary arithmetic is important because of computers. This is not strictly true. Binary arithmetic is important because working in this system makes children think about the basic principles of arithmetic and so helps their understanding of everyday arithmetic and hence improves their skill and accuracy. While some computers work in binary, others do not; it is not possible or necessary to tell which is which from the operating position.

The use of symbols and shorthand

As noted above, the symbols that we use for numbers are called numerals. For the number of dots following: . . . . we use the numeral 7, and for the number which multiplies the diameter of a circle to give its circumference we use the numeral π. These are the nouns of mathematical language. The verbs are the symbols for relationships, Usually the first relationships that a child will meet are 'greater than' and 'less than', applied concretely, 'Bill is taller than Joan.' Next is 'is equal to'. A child must be able eventually to translate 'seven twos are fourteen' or 'seven times two is equal to fourteen' into the shorthand '$7 \times 2 = 14$' and 'eight is greater than seven' into $8 > 7$. Just as the child can turn around an English sentence, and express the same thing in various ways, so he must be able to alter this mathematical sentence. He can write, $2 \times 7 = 14$, or $14 = 2 \times 7$, and so on. This process is similar to taking a simple sentence like 'I am going home', and re-phrasing it as 'I go home', 'Home is where I am going', and so on. Sometimes shades of meaning occur in English sentences, but, in general, mathematical relationships mean the same thing any way round.

The formal writing of these relations depend on age and ability. When he is ready a child should 'play' with symbols and numerals, which are a sort of abstract apparatus, in the same way as he plays with concrete apparatus. He will write:

$$4 + 3 = 7,$$
$$\text{also } 4 = 7 - 3,$$
$$\text{also } 3 = 7 - 4,$$
$$\text{also } 4 + 3 - 7 = 0.$$

He is forming the 'number bonds' between the three numbers 4, 3 and 7. These lead him into simple problems like these, in which it is necessary to find the right number to fill the query boxes:

$$4 + 3 = \boxed{?}$$
$$4 + \boxed{?} = 7$$
$$7 - \boxed{?} = 4$$
$$\boxed{?} + \boxed{?} = 7$$

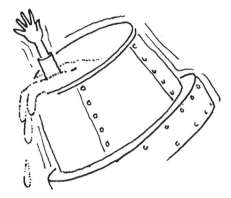

'in the same way as he plays with concrete apparatus'

Open-ended questions

This last problem is an interesting one, since it is 'open-ended', that is to say, there is no single correct answer, but an indefinite number of correct ones. It is typical of the new way in which mathematics is being taught that it has a creative side from the word 'go'. Children can find interest in being asked to invent several calculations to which the answer is 7. A primary school class responded with great enthusiasm to the suggestion that they should find out all they could about the number 12. Just a few of the things they wrote are as follows:
12 is:

> an even number between 11 and 13
> spelled with 6 letters, which is half of 12
> the last age before a teenager
> the number of Jacob's sons, disciples, people on a jury
> sides to a 3d bit in pre-decimal coinage, inches in a foot
> months in the year, days of Christmas
> a rectangular number: 2, 3, 4 and 6 divide it
> 12 August is the opening of the grouse-shooting season
> Columbus discovered America on 12 October 1492
> 12 × 12 yards is the normal pace for archery

What splendid research!

Language of algebra

Many primary schools also introduce the rudiments of the language of algebra. The number of units of area in a rectangle is found by multiplying together the number of units of length in its length and the number of units of length in its breadth. More shortly, the area of a rectangle is found by multiplying its length by its breadth.

More shortly still, $A = l \times b$. Once the meanings of these letters have been clearly defined, the statement can be put down very shortly. This is an example of a formula. Another formula says $C = \pi \times d$. This expresses in shorthand form the relationship noted above, that the circumference of a circle is obtained by multiplying its diameter by the number π. Older children find out how to turn English sentences into algebraic symbols and also to translate a formula back into an English sentence. The formula $t = 20 + 45 \times W$ may be understood clearly when we are told that W is the weight† of a joint of meat in kilograms and t is the time of cooking in minutes. The formula can then be read as 'The time of cooking of a joint of meat is 20 minutes and 45 minutes for each kilogram weight of the joint.'

Mathematics from everyday life

The mathematics in the primary school does not spring entirely from a specially structured world, of course. Children are encouraged to be interested in the numerical aspects of things around them, for instance, the traffic on the roads, rainfall, quantities of milk or chocolates consumed by their respective classes. Much of this will involve graphing information. The graphs that they draw will not just be spikey straight lines, like the humorous temperature chart so often drawn by the cartoonist at the end of the invalid's bed. Children are encouraged to invent their own methods of representation. They like to use pictograms in which the numbers to be pictured

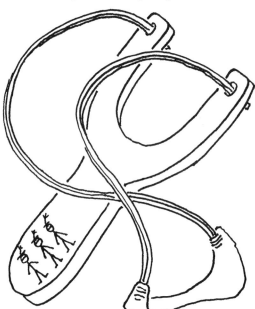

'They like to use pictograms'

† See Answer 42 at the back of the book.

are shown as proportional to the numbers of little characters, children, bottles of milk, etc. An example of a graph drawn by a primary school child is shown in Fig. 3.15(*a*).

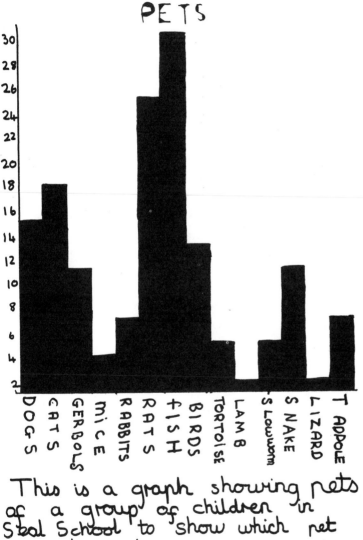

Fig. 3.15(a)

Pie chart of favourite types of television programmes.

The pie chart shows, that the majority of children like comedy best. Serial is the second choice followed by cartoon. Nature and westerns are the least popular.

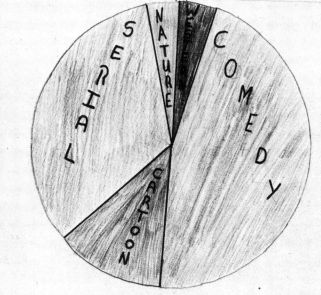

Types of programmes	NO.	Angle
Western	1	15°
Comedy	11	165°
Cartoon	3	45°
Serial	8	120°
Nature	1	15°

The 360° angle at the centre is divided into 24 part of 15°.

Fig. 3.15(b)

They will also draw pie-charts (Fig. 3.15(*b*)). This type of graph calls for an understanding of proportionality and of angle. Here is an example. In a certain week there were 72 hours of broadcasting from a television station. This was made up as follows:

	Hours
Drama	6
Films	15
News and documentaries	8
Children's programmes	10
Music	12
Comedy	9
Other	12
Total	72

The angle at the centre of the 'pie', 360°, will be split in the same proportion.

Since 360° corresponds to 72 hours
then 5° corresponds to 1 hour

and so we get the following table of angles:

Drama	30°
Films	75°
News and documentaries	40°
Children's programmes	50°
Music	60°
Comedy	45°
Other	60°
Total	360°

(1 hour ↔ 5°, so 6 hours ↔ 30°)

This produces a diagram as shown in Fig. 3.15(*b*). From this the relationship between the times allocated to the various types of programme is easy to see. Furthermore, if another pie chart is drawn for a different week, it will be easy to compare the way in which the time given to any particular type of programme has changed. At this level it is necessary for the teacher to arrange for there to be a simple relationship between the hours and the angles, for instance, 1 hour to 5° or 1 hour to 10°. Real life presents more difficult calculations, but the reasoning is the same; the arithmetic may be beyond the child's ability but the concepts are not.

Apart from graphs drawn by children, time is spent on discussing how to understand information presented to them graphically by others. Advertisements, government publications, newspaper articles and books contain plenty of examples. Even in the primary school it is easy for children to learn to be critical, to object to statements such as 'twice as white', or to graphs with no scales shown (see Chapter 10).

The place of drill

The place of drill or rote-learning has to be carefully re-considered. Once a concept (for example, the concept of multiplication) has been well understood, there is great value to the child in knowing quickly and accurately the multiplication tables. Very many children, however, find learning these both hard and unattractive. It is considered, rightly we think, that it is a mistake to hold up a child's progress in mathematics simply because he or she is not able to memorise tables. However, every child should be encouraged to do the best he can. If he has difficulty, most schools will now allow a child to have printed tables and to refer to these while doing sums. Many schools are experimenting with the use of calculating aids, such as ready reckoners, hand calculating machines, or even pocket electronic computers. These are particularly valuable in the case of the backward child who never has confidence in his ability to calculate and whose progress need now not be halted.

Rote-learning of methods is hard to justify. It is certainly true that one of the hardest parts of mathematics is to understand the nature of the problem, that is, to translate the facts and questions into the proper mathematical form so that calculation or reasoning can start. There is remarkably little educational value, however, in making this difficulty the excuse for rule-of-thumb methods of the type, 'When you get a problem like this, you deal with it like that.' Satisfactory learning has a transfer value; that is to say, a method which works for one type of problem can be modified and made to solve a different one. Unless the method is understood, such transferability will not exist.

In this chapter we have mentioned only a small selection of the topics met by children in the primary school. There are many others; sets, statistics, symmetry, topology to name but a few, and some of these are touched on later, in chapters on the secondary stage of mathematical education.

Some parents of today's primary school children complain that their children know less than they themselves did. 'At their age,' they will say, 'I could do all my tables and understand simple interest.' Such criticism of the new primary courses is ill-informed. At this stage of the child's development it is not the amount but the depth of his understanding and knowledge which is important. In the primary schools the foundations are being made for all future progress, and a concept not thoroughly grasped at this stage may well be the cause of a complete blockage when the child is 14 or 15.

Of course, it is true that providing the structured surroundings and then allowing a child to discover results, to plan its own work, to play with concrete material, takes time, but it is time well spent if, as a result, a child has a sound basis for future work and also leaves the primary school with a love for the subject. Good foundations always take time to build, but woe betide the structure, whether of bricks or of knowledge, which is without them!

4 Changes in secondary school philosophy

How they started

Once the rumblings of change in primary schools had erupted into action in the late 1950s, it became inevitable that secondary school mathematics would have to change too. Not that the seeds of revolution were not growing there too. Mathematics teachers in secondary schools were all too aware that mathematics was broadly disliked as a subject at school, to the extent that grave doubts were being expressed whether enough mathematicians were being produced to continue to man the schools. This was put most forcibly by Bryan Thwaites in his inaugural lecture as Professor of Applied Mathematics in the University of Southampton. It was then that SMP found its prophet and Director. A few individuals had been experimenting with their own syllabuses before that date, but this speech in 1961 gave birth to the first project to make a national impact.

With some help from university teachers, inspectors and mathematicians in industry, groups of young teachers came together in various parts of the United Kingdom to write new material for their schools and to look for ways to rejuvenate old material. We were lucky in being able to read accounts of earlier experimentation in the USA. Although we adopted a different line, we found their work stimulating and provocative. Much of the new material had never been taught to children before. It was therefore essential to arrange for experimental classes to test out its suitability and impact. So in the early 1960s classes of children in a number of schools suffered the frustration, but also the excitement, of arriving each Monday morning to see what had been written for them over the weekend. They worked from duplicated sheets, full of typing errors and slips in working. A number of the happier touches in the texts eventually printed owe their origin to the classroom discussion that followed, or to the quick response of a child. For teachers other than the writers this was also a time for great activity, as they asked for copies of the rough drafts and tried to master their contents, to see whether they agreed with what was being produced and to see if they wished to associate their own schools with the movement.

It is one of the strengths of the school system in the United Kingdom that it is possible to change syllabuses freely, and to have these changes reflected in new public examinations. Any school which wishes to be examined on its own syllabus can submit it to the Schools Council. Specialists will examine the syllabus in detail to make sure that it measures up to proper standards, not too hard, not too easy,

[44]

and if it does, it will be accepted. The discipline of having to write an examination syllabus with specimen questions and, later, to set examination papers helped to give definition to the new ideas.

Areas of change

Today there are hosts of modern textbooks and materials for children of all abilities and the teacher faces a very real problem of selection. It is interesting to note, however, that they broadly agree in the direction of change. Although we shall analyse some of the changes in detail in later chapters, it may be helpful to give a broad summary of what has changed.

1. Courses have become more algebraic and algebras other than the algebra of numbers have been introduced. They comprise sets, geometrical transformations, matrices and vectors, all of which tend to be studied from the algebraic standpoint at least in part.
2. They stress the ideas of relation and function which are central to all mathematics.
3. Euclidean geometry, with its theorems, riders and constructions is played down or has virtually disappeared. Children are encouraged to carry out experiments in geometry, both two-dimensional and three-dimensional, and ideas of measure, area and volume are derived from transformations.
4. Statistics and probability are introduced.
5. Logarithms give way to the slide rule as the main computing aid, but calculating machines, mechanical and electronic are not ignored. The place of the computer in modern society is stressed and some courses include detailed programming.
6. The idea of a mathematical model of a physical situation pervades applications of the subject.
7. Emphasis is more on understanding and less on memorisation and mechanical responses.

Planning the curriculum

It has already been noted that much of 'new' mathematics is not new. Further, there is equally no implication that 'old' mathematics is necessarily bad or unsuitable. What is aimed at is a synthesis of the best of both. When deciding whether or not to include a topic there are three main things to consider as regards its content. First, we consider the help that the topic gives in building up the fundamental structure of the subject, whether it leads on to further important topics and whether it links with other ones. Second, we consider whether it is good mathematics, thought-provoking, intriguing, containing the unexpected, elegant. Then, third, we consider its applicability to everyday life and to the needs of other subjects. Mathematics must not be taught as a bag of tricks. It has unity, philosophy and a language of its own, as well as utility. What is more, to practise mathematics should also involve aesthetic appreciation. (Is this method more satisfying than that?) At the heart of it all is what is called generality.

Generality

It is easiest to explain this by way of examples. We shall look at three concerning prime numbers. Men have long been fascinated by the primes, that is the numbers like 7, 11 and 101 which have no divisors or factors other than themselves and, of course, 1. The first few, in order, are as follows: 2, 3, 5, 7, 11, 13, 17, 19. As you can see they lie irregularly on a number line (Fig. 4.1). They have been underlined.

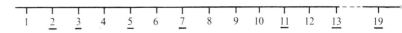

Fig. 4.1

Many attempts have been made to see just where they occur and to be able to predict which the prime numbers would be between, say, 1 000 000 and 1 000 019. Men have studied how they can be expressed as combinations of other prime numbers, not as multiples, of course, since by definition a prime has no factors, but perhaps as the result of additions or subtractions.

We may see, for example, whether a prime can be expressed as the sum of two smaller primes. Here are the first few, which certainly can:

$$2 = 1 + 1; 3 = 2 + 1; 5 = 2 + 3; 7 = 2 + 5.$$

What do you find with the next one? What do you find if you try to express the first few primes as the difference between two other primes? The result underlines one fact about the primes which should be obvious but somehow never quite is! Look in the back if you are stuck (Answer 3). Interestingly, it is enough to find one case of failure to disprove a whole theory such as 'Every prime can be expressed as the sum of two smaller primes.' How many successes would be needed to prove that such a theory is true? (See Answer 4.) We shall return to this question at the end of the book (Chapter 13).

Our last investigation concerning primes involves the use of a formula to try to 'generate' primes. Here are the first few lines of the formula written out numerically

$$1^2 + 1 + 41 = 43 \text{ which is prime};$$
$$2^2 + 2 + 41 = 47 \text{ which is prime};$$
$$3^2 + 3 + 41 = 53 \text{ which is prime}.$$

If you continue this series, working out each one, you will find that it continues to give prime numbers, probably until your patience gives out. But will it always?

Let us look at the formula expressed algebraically. It is

$$n^2 + n + 41,$$

where n is any positive whole number. Written like this it is not hard to see what will happen if we put $n = 41$. The formula will give:

$$41^2 + 41 + 41.$$

There can be little doubt that this has a factor, 41 itself, and so the formula is here generating a number which is not prime. You may be able to find a smaller value for n which yields a non-prime, but there is no need for this, one failure disposes of the conjecture and a disproved theory is of no interest! The formula $n^2 - n + 41$ also gives a long string of primes, but falls down at least for the same value. Can you name another value for n for which both formulas will certainly fail to give primes? (See Answer 5.)

It is now tempting to try other algebraic formulas, for example $4^n + 3$, but this too breaks down. (See Answer 6.) In fact there is no known formula for generating primes algebraically in this manner.

The sequence of ideas that we have been through is typical of mathematical generalisation. In the first place we look at particular number relations. Then we try to look for a relation which fits, not just a few, but all numbers. Then we find ourselves contemplating a number of different formulas and classifying them in some manner, for example, as algebraic.

Here is a non-algebraic way of generating a prime: take the first n primes, say twenty, multiply them all together and add 1. This number is definitely prime, can you see why? However as there is no algebraic way of writing out the first n primes there is no algebraic formula for this number. As a matter of interest, however, this does demonstrate that there is no largest prime since this construction, given all the primes you can think of, will always provide a larger. Thus primes 'never stop'.

Each generation of mathematicians tries to extend and bind together the apparently unrelated results of its predecessors. This generalisation process is one of the main-springs of mathematical research. If it succeeds the mathematics gets simpler to look at — but deeper to understand! However, for each generation of children mathematics is fresh, truly waiting to be discovered. It is for the teacher to keep abreast of new ideas and to act as guide.

In addition to the three criteria for content mentioned above, the teacher has to add two more. He must consider the suitability of each topic for learning at any particular age. It may be too abstract, too difficult a concept for the child's mental development (compare the concept of conservation touched on in Chapter 3) or require too much manipulative ability or skill in calculating. He also has to consider motivation.

Motivation

It is a truism that children do well at what they like and do badly at what they do not like. When the child is young, liking for a subject is often associated as much with the personality of the teacher as with the actual content of the work. As the

'associated with the personality of the teacher'

child grows older he usually learns to dissociate the subject from the person. Then, of course, liking reinforces the doing well and disliking the doing badly so that a 'blockage' can be built up at quite an early age and used as an alibi to excuse subsequent lack of effort.

Far too little research has yet been done to find out what boys and girls consider, at any given age, to be important in mathematics. Some remarks made by thirteen-year-old boys in the early, testing, days of SMP may throw some light on this.

'what boys and girls . . . consider to be important'

One wrote, 'My best bit is probability, because it is useful. I don't like matrices so much. They are easy. Who uses them?' This was in answer to an invitation to criticise the new work that he had been given and to say which bits he liked, and if possible, why. Another boy wrote, 'It is much more interesting for those who want to continue with mathematics after O-level.' Another, 'What will happen if it is a flop subject and employers refuse to recognise it?' And yet another, 'At my old school I thought mathematics was one of my bad subjects, but now I find it extremely interesting and useful.' Even allowing for a tendency to write what the teacher wanted to hear, this emphasis on the utility of the subjects, either for everyday life, or for further education, or in the hunt for jobs, is interesting. It has implications for a writer to know that boys of this age tend to be so material-minded.

On the other hand, younger boys, eleven-year-olds, selected some of the more abstract parts of the work as being most interesting to them. Binary arithmetic figured highly on their lists, and this is far from being a utilitarian topic, as we have already noted.

In addition to what they think of as relevance to everyday life, many children like to be able to be practically involved, to do experiments. These may either be conducted in the world about them, for instance a traffic census or the measurement of times of swing of pendulums of varying lengths, or the experimentation may be with numbers, or mathematical shapes, or other abstract ideas. To illustrate their work children like to make drawings or graphs, preferably in bright colours, and the modern mathematics classroom is not normally short of material for murals.

It must not be forgotten, in this concentration on the practical, that there is also plenty of motivation in unexpected, elegant or paradoxical results. Children of all levels of ability like to find mistakes, not in their own work, but in the work of others. A sheet of worked examples plentifully sprinkled with errors in calculation, and flaws in arguments usually finds plenty of takers.

None of this motivation is particularly new. The best teachers have always made use of these ideas. What is new is that many of the 'fun' ideas, which had to be fitted in when time allowed, are now part of the syllabus. This positive motivation is such that, when the subject-matter is interesting and exciting, it becomes its own reward. It contrasts strongly with the negative motivation embodied in fear of punishment if you get the answers wrong.

Feedback

It has been discovered that children react in the way that they are expected to react. In an experiment in the USA some slow learners were given a test which was, in fact, completely meaningless. However, half the children who had taken it were told

'a test which was, in fact, completely meaningless'

that they had done really well, that the test results showed that they were much better at the subject than anyone had previously suspected. The other half were told that the results merely confirmed existing estimates of their ability. In the next piece of work that the children handed in, those who had been told they were better than they thought, did better, while those who had been told nothing, did as well as usual.

That feedback of this type exists will not surprise experienced teachers, though it is always good to have one's intuition confirmed by experiment. We have long known that we can 'jolly along' children, given time and the right environment. It goes a long way to explain the success of the 'crammer' where children who have failed vital examinations are sometimes sent to be 'crammed' for a second go. The combination of individual tuition, a business-like atmosphere ('Over 80% success . . . ') and the expectation that crammers get results often means that they do get results.

It is this knowledge, now strengthened by experiments of the type outlined above, which has led teachers to think very hard about the effect of awarding marks or grades. Low marks can confirm a child in his own low estimate of his own ability and lead to a continuation of the same or worse. Various suggestions have been made for avoiding this, and some schools have adopted these or similar ideas. One suggestion is to avoid any form of numerical assessment and rely solely on verbal comments such as 'You are doing really well!' or 'Certainly some improvement on last week!' Another is to provide children with worked solutions before any mark is awarded, so that they can spot their mistakes and always hand in correct answers. When children are working together, in pairs or larger groups, assessment is sometimes made only of the total effort, not of any individual's contribution.

There are snags in all these. Children seem to like to be assessed; they like to know how they are doing, even if it is not very well. They are not taken in by a pretence that they are not being judged, and sometimes find methods of judging themselves. Then again, it is plainly necessary for teachers to have records of progress, and parents tend to get restless if teachers refuse to commit themselves to any view on how individual members of their class are doing. As long as public examinations exist there is also the point that children need practice in taking tests and examinations. So despite theoretical worries most writers continue to write tests into their books or their material and a good many parents continue to be supplied with 'Fortnightly Orders' or similar lists of marks or grades relating to their child.

Well-judged tests, humanely marked and with low scores softened by suitable remarks, remain part of the motivational armoury of most teachers. But the prime motivator remains the teacher in the classroom. If he or she has a real enthusiasm for the subject this will infect the class too. Often a new syllabus will be taught better than an old syllabus, just because it is new and has involved the teacher in thinking out new approaches and in justifying to himself the work that is going to be done. The early days of a new syllabus are usually encouraging, but before assessment of it is meaningful, some time must be allowed to pass. (We return to this in the final chapter of this book.) If the new syllabus material really is superior to the old (and we have no doubts about the new mathematics!), there is thus an added bonus.

Transition from primary school to secondary school

Moving from the primary to secondary level (at about 11 years old) has always posed difficulties for the average child. The small school is replaced by a large one, the class teacher by a whole range of subject teachers and often in the past the style of teaching has changed too. The problems have not been confined to mathematics, of course, but as already remarked, new ideas in primary school mathematics teaching generally preceded changes in the secondary schools, and so the effects were particularly severe. It is easy to upset a child's newly awakened interest in the subject by insisting too soon and too often on traditional layout, rules or drill. The transition will be accomplished most easily when the secondary school continues with project work, or with work in groups, and where modern syllabuses have been adopted which form a natural continuation of primary school work. As time goes by the child must learn to be more formal and to accept that some work may be repetitive, or even dull.

'the child must learn to be more formal'

It is an encouraging sign of the times that primary and secondary teachers are coming together much more to share information. The Teachers' Centres now to be found in most parts of the country often provide the setting. The major projects (such as SMP), the Department of Education and Science and many local colleges and institutes of education run courses and provide opportunities for in-service training. It is to be hoped that the proposals of the James Report will be put into effect before very long. The most significant of these was that a teacher should be released, after every seven years of teaching, for refresher courses or further study. With the present rate of change in mathematics syllabuses this innovation would be of great significance for the full flowering of the modern movement.

Middle schools

One way out of the difficulties of transfer at age 11 is to change the date of transfer. For various reasons, including the desire to keep schools reasonably small and to make the best use of existing buildings, there has been a dramatic growth in the number of 'middle schools' during the past ten years. A middle school caters for the age range 8–12, or 9–13 normally. It has been said that, by 1980, as many as a quarter of 11-year-olds may be in middle schools. Inasmuch as middle schools represent a sort of half-way type of school between primary and secondary, it may be supposed that transfers may be easier, but surely problems will remain! It is still early days for these exciting new schools (although the private ones, the so-called preparatory schools, have been with us for many years) and we shall not deal in this book with the mathematical aspects of these problems.

What sort of secondary school children benefit from new mathematics?

Most of the secondary projects started in grammar schools or public schools (private schools). The main reason was that it was in these schools that most of the trained mathematics graduates taught. The new material that they wrote was therefore aimed at the top third or so of a total 'ability range' from weakest to brightest, though children were selected for these schools on general ability and not specifically because they were good at mathematics. It is interesting to study the valuable list of these projects which has been published for the Mathematical Association and is available through bookshops (*Mathematics Projects in British Secondary Schools*, published by G. Bell & Sons, second edition in course of preparation). For each project it gives details of the professional staff involved, a brief history, and details of the age and ability range of the children aimed at. In the back is a detailed comparison of the syllabuses of the nine largest projects.

Initially there were some doubts expressed as to the suitability of modern syllabuses for the exceptionally gifted child. It was feared that he might not have made as much progress as the traditionally trained child by the time he left school. Experience has shown that, while there are some areas which are no longer covered, these children are better able to cope with modern higher education courses in mathematics and, if less skilled in certain techniques, possess the resilience and confidence quickly to pick up those necessary for courses in science, engineering, etc. Of course it is necessary for lecturers in universities and polytechnics to recognise that a change has taken place and to respond to it.

For the able child, not in this gifted class, modern syllabuses have proved themselves to be admirably suited. The evidence for this comes from the ever increasing number of children taking mathematics to A-level in the sixth form. Not all of these will also be taking science courses. It is common for mathematics to accompany arts subjects as well as new subjects like economics, design, sociology and business studies.

Soon after the original draft texts of the larger projects had appeared the teachers involved began to see the possibility of using the same topics, in a more simplified form, for the average or below average child too. This was something of a surprise, for despite the concrete approach, much of the material is more abstract and 'mathematical' than traditional work. Soon lower streams in grammar schools and children in secondary modern schools and comprehensive schools were working from re-written texts and the earliest work-cards were beginning to appear. Now that the unstreamed comprehensive school is becoming common, the methods described in Chapter 2 are coming more and more to the fore, and modern mathematics is very far indeed from being the preserve of the brightest. In fact, it is found that some of the modern topics are most suitable for the so-called 'remedial' classes, that is for those children whose ability is so low that they have considerable difficulty in coping with the concept of number.

Virtues of the old mathematics

As you may have gathered from the occasional remark, the adoption of the new syllabuses did not occur entirely without rearguard action from traditionalists. No

one would deny that there is much good stuff in traditional syllabuses. Indeed the new ones are only perhaps 30—40% new in content, though broadly different in spirit, and there has always been scope for the gifted teacher to inspire and motivate. As the new work started to spread, some people expressed concern about the mental training that is bestowed, and deplored in particular the disappearance (or virtual disappearance) of Euclid's geometry. More will be said about this later, and in Chapter 7 the nature of this change will be explained. The fact is that there is considerable doubt whether the formal training in logic that was supposed to be the outcome of a geometry course based on Euclid really had much transfer value. Does knowledge of Euclid's theorems and ability to work problems on them have much effect on a child's ability to argue logically about politics, ethics, religion, etc.?

'ability to argue logically about politics'

Concern was also felt lest the child brought up on modern mathematics should be dependent on machines to calculate for him, partly because mechanical aids were being held up as good in themselves and partly because he was not being sufficiently practised in computation. It is probably true that, in the early days, the new parts of the syllabus loomed over-large in teachers' minds, and basic arithmetic was not adequately rehearsed. It is hard to substantiate this, however. There is still debate, but teachers are now familiar enough with the new courses to keep a proper balance between initiative and exploration on the one hand and skill and accuracy on the other. It must not be forgotten that each of these wings of mathematical education is equally important and neither can exist without the other.

The time has come to turn our attention to the details of the new courses. The following chapters continue to be about secondary school mathematics, but it has seemed best to divide them up into subject areas: algebra, geometry, and so on. In order to understand more fully what your child is doing we are inviting you to do more of the mathematics for yourself. Each chapter starts relatively easy and gets harder. If you find yourself getting out of your depth, despite the answers and notes in the back, we suggest that you stop and move on to the next chapter. When you have finished Chapters 5—12 you may like to return to this chapter and re-read it in the light of your greater experience. The only way to grasp mathematics is to do it! Please remember that it is harder for you, with your own schooling probably traditional and, perhaps, some years past, than it is for your child with a young, fresh, uncluttered mind! However, we hope to convince you that modern secondary courses do result in more mathematics, better mathematics and mathematics for all.

5 Algebras

When you saw the 's' on the end of the word 'algebras', did it cross your mind that this might be a printer's error? Most of us met with only one sort of algebra in our school-days and it concerned xs and ys which stood for numbers. But the algebra of numbers is only one possibility. There are other algebras, each with its own set of rules, and the rules themselves depend on the nature of the *elements* being manipulated. Manipulations are called *operations*, and signs like = which connect them are called *relations*. Familiar operations like addition, subtraction, multiplication and division may still occur (there will be new ones too), but have to be differently defined. We shall look at three simple algebras, odd and even, circuits and sets. The last is the most used but, as it may be least familiar, we will leave it to the end of the chapter.

The algebra of odd and even

Write down any number that occurs to you: it will either be divisible by 2 or it will not. If it is divisible by two without remainder it is even: 2, 4, 6, . . . are even numbers; if not, it is odd: 1, 3, 5, 7, . . . are odd numbers. We call all the even numbers one 'class', and all the odd numbers another 'class'. Any number belonging to the class of even numbers will be referred to as E and any number in the class of odd numbers will be referred to as O. It is one of the characteristics of an algebra that we start by giving letters to the mathematical abstractions involved; this serves as a sort of shorthand. If we take two numbers from the even class and add them, to which class will our answer belong? Consider some examples: $8 + 6 = 14$, $10 + 12 = 22$. An even number added to an even number always gives an even number, or more neatly, we write $E + E = E$. Now consider an odd number added to another odd number, for example, $1 + 3 = ?$, $5 + 7 = ?$ Now complete the statement $O + O = ?$ If we add an odd number of an even number, to which class does the answer belong? Try $1 + 4$ and $11 + 6$, for example. Now complete the statement $O + E = ?$ Would it make any difference if we had started with an even number and added an odd number? Complete the statement $E + O = ?$ (See Answer 7 at the back of the book.)

Commutative operations

If we get the same outcome regardless of the order in which we combine two things, the operation which combines them is said to be *commutative*. The operation calle division is not commutative since $12 \div 4$ is not equal to $4 \div 12$. Most everyday actions are not commutative when performed. Consider the acts of putting on soc and putting on shoes. Here the order certainly does matter when we combine then by the operation 'followed by'! 'Put on shoes' followed by 'Put on socks' is not th same as 'Put on socks' followed by 'Put on shoes'. 'Followed by' is a non-commutative operation on these actions. A convenient way to summarise our resu for the operation of addition on O and E is to make a table as in Fig. 5.1 In the to

+	O	E
O	E	O
E	O	E

Fig. 5.1

left-hand corner, the + sign shows the binary operation we are performing. It is called binary because we combine two things at any one time (compare bilingual, biped, bicycle, etc.). To use the table to find the 'answer' to $O + O$, start on the left-hand side of the table, run your finger down until you find O, then run along the row stopping in the column headed O, getting $O + O = E$. Similarly $O + E = ($ and $E + O = O$.

×	1	2	3
1		2	
2			6
3			

Fig. 5.2

×	O	E
O		
E		

Fig. 5.3

Tables are important in this kind of work. If you would like to try two more, look at Figs. 5.2 and 5.3. The first is about numbers and the operation is multipli-cation. Can you complete it? The 2 in the body of the table was obtained by multi plying the 1 underneath the multiplication sign by the 2 in the top row by the side of the multiplication sign. The completed table looks like Fig. 5.4.

Another you might like to try concerns odd and even numbers under multipli-cation; Fig. 5.3 gives the outline and you can check your table again by looking at Fig. 5.5.

×	1	2	3
1	1	2	3
2	2	4	6
3	3	6	9

Fig. 5.4

×	O	E
O	O	E
E	E	E

Fig. 5.5

These tables summarise our calculations with some binary operations. They can be used to simplify expressions containing more than two terms. We can use Fig. 5.1 to simplify O + O + E + E + O. As the table only tells you the result of combining two elements at a time, it is first necessary to pair off the elements, for instance, like this:

$$O + O + E + E + O = (O + O) + (E + E) + O.$$

Now, from the table in Fig. 5.1, we see that O + O = E and E + E = E, so the simplification can be completed like this:

$$
\begin{aligned}
O + O + E + E + O &= (O + O) + (E + E) + O \\
&= \quad E \quad + \quad E \quad + O \\
&= (E + E) + O \\
&= \quad E \quad + O \\
&= O.
\end{aligned}
$$

The brackets 'punctuate' the statement in the same way that commas punctuate a written sentence, and just as by putting a comma in a different place the meaning of a sentence can be altered, so can the meaning of a mathematical sentence. If a

'*a mathematical sentence*'

newspaper headline said of a football match, 'France routed England win', does it mean 'France routed England, win', or is it 'France routed, England win'? Would it matter if in working out $O + O + E + E + O$ the brackets had been put in like this: $O + (O + E) + (E + O)$? Working it out again,

$$O + O + E + E + O = O + (O + E) + (E + O)$$
$$= O + O + O$$
$$= O + (O + O)$$
$$= O + E$$
$$= O.$$

In this example it makes no difference. By reasoning, or by looking at every possible combination, we can see that this will always be so for this operation and with this set of elements O and E.

Associative operations

When it does not matter where the brackets are put we say that the binary operation we are using is *associative*.

Try using the same table to work out:

(*a*) $E + O + O + E + O + O$;
(*b*) $O + O + O + E + E + E$;
(*c*) $O + E + E + E + O + E + O$.

(See Answer 8 at the back of the book.)

Children are introduced to the algebra of O and E when quite young and also to the ideas of commutativity and associativity since understanding of these is a help in ordinary arithmetic of numbers. It is true that $6 + 3 = 9$, and $3 + 6 = 9$, etc., the commutative rule holds for numbers under addition. The associative rule also holds for numbers under addition. Consider $5 + 4 + 6$: we are so used to working this in our heads that it takes a positive effort to slow down and start by grouping

$$(5 + 4) + 6 = 9 + 6$$
$$= 15.$$

Again,

$$5 + (4 + 6) = 5 + 10$$
$$= 15.$$

Working one example proves nothing, of course, but it is not hard to see that $x + y = y + x$ and $x + (y + z) = (x + y) + z$ for all numbers x, y and z.

What about other operations? Is subtraction commutative and associative? Is $8 - 4$ the same as $4 - 8$? Does it matter if I have £8 and a bill for £4 or if I have £4 and a bill for £8? It certainly does! Subtraction is not commutative. One example is enough to show this.

'Does it matter if I have . . . £4 and a bill for £8?'

Now consider $15 - 4 - 3$.

$$15 - (4 - 3) = 15 - 1$$
$$= 14.$$
$$(15 - 4) - 3 = 11 - 3$$
$$= 8.$$

Here it does matter where the brackets are placed, the operation of subtraction is not associative, we have to make a rule about 'changing the sign' when we deal with subtractions.

This is not 'new' mathematics but present-day teaching methods stress the need to establish a clear understanding of the rules which can be applied in any given situation, and the rules should be 'discovered' by the children themselves working suitable simple examples. A natural outcome of this is that words like 'commutative' and 'associative', referring to these rules, are now in common use in school mathematics lessons. More importantly children should use these rules where appropriate. Find quick methods for these:

(a) $997 + 1998 + 3$;
(b) $\frac{1}{4} \times \frac{1}{3} \times 20 \times 30$.

(See Answer 9.)

You may ask what use is the algebra of O and E? It is not useful in the most immediate sense. But quite young children can build up the table and in doing so they have to think about numbers, and in using the table to simplify such things as $O + E + O + O$, have gained experience in handling symbols. The table can be used to solve equations, too, just as in any other algebra. An equation makes a statement about equality. If we write $O + ? = O$, we are asking 'What, under the operation of addition, combines with O to make O?' We have only an O or an E to choose from. From the table we see that $O + O = E$, so it cannot be O. If we try E, then from the table we see that O + E = O, so the answer is E. We could have used the conventional letter x to stand 'for the quantity to be discovered'. Then we would have written O + x = O. If we first tried replacing x by O, we should have O + O = O, which is not true, and if we then tried replacing x by E, we should have O + E = C which is true. So x = E.

Closure

Another idea in the structure of algebra is that of *closure*. Under addition the algebra of O and E is *closed*. This means that the only elements that occur in the table are O and E themselves. In contrast, let us make up the addition table for the numbers 0 and 2 (Fig. 5.6).

+	0	2
0	0	
2		

Fig. 5.6

0 + 0 = 0, and this has been entered. Fill in the remaining spaces in the table (see Answer 10). This is easy enough, but when we come to 2 + 2 we have to use a number other than the 0 and 2 with which we started. So we do not find a closed algebra of 0 and 2, under addition.

The algebra of electric circuits

If electric circuits are outside your sphere of interest, do not give up, for this algebra can also arise from consideration of water flowing through a suitable arrangement of pipes, and indeed from many other things! You may know all about electric circuits, in which case please skip the next few sentences; on the other hand, you may not know about them but believe (as you should) that what young children can cope with is not beyond your understanding. For you the following explanatio of the symbols used in the diagram of an electrical circuit will be necessary (Fig. 5.7). The symbol ⊩ near the letter C indicates a battery; A and B are switches Electricity can only flow and light the lamp L if both switches are pressed down so that there are no gaps in the circuit. If A is pressed down and the gap is closed, current could flow through A, and we shall use the number 1 as shorthand to mean 'current flows'. If, at a switch, there is a gap and no current can flow, the number 0 will be used to indicate 'no flow'. If current is flowing through the lamp, it will light up and we shall write 1; no light is shown by 0.

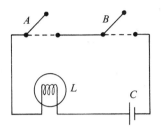

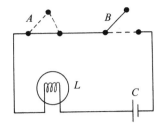

ig. 5.7   *Fig. 5.8*

For Fig. 5.7: $\left(\begin{array}{c}\text{Switch } A\\ \text{off}\end{array}\right)$ and $\left(\begin{array}{c}\text{Switch } B\\ \text{off}\end{array}\right)$ results in $\left(\begin{array}{c}\text{Lamp } L\\ \text{off}\end{array}\right)$;

or in symbols: 0   0   =   0.

For Fig. 5.8: $\left(\begin{array}{c}\text{Switch } A\\ \text{on}\end{array}\right)$ and $\left(\begin{array}{c}\text{Switch } B\\ \text{off}\end{array}\right)$ results in $\left(\begin{array}{c}\text{Lamp } L\\ \text{off}\end{array}\right)$;

or: 1   0   =   0

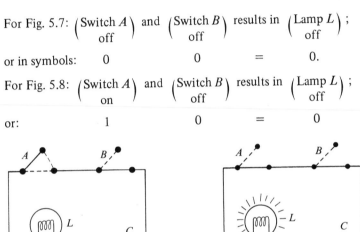

Fig. 5.9   *Fig. 5.10*

Abbreviating for simplicity we also have:

For Fig. 5.9: off and on results in off
or: 0 + 1   =   0.

For Fig. 5.10: on and on results in on
or: 1 + 1   =   1.

The table in Fig. 5.11 summarises these results.

			Switch B	
			On	Off
		+	1	0
Switch A	On	1	1	0
	Off	0	0	0

Fig. 5.11

For those who would feel more at ease with water than with electricity, all tha·
is necessary is an imaginary supply of water and a pipe leading out of it, in which
there are two valves (or taps) A and B, to control the flow of water as in Fig. 5.12
If a valve is in an open position allowing water to flow through, we shall put a 1,
and we shall use 0 to show that no water can pass when the valve is closed. If the
manipulation of the valves allows water to flow from the end of the pipe, this too
will be denoted by 1 and no flow by 0. The diagrams need little explanation. Fig.
5.13 summarises these results, and comparing it with Fig 5.11, we are not surprise
to see that it has the same pattern.

In the electric circuit and in the water pipe, the control mechanism B followed
control A directly, and with such an arrangement the switches and valves are said t
be 'in series'. In the two diagrams in Fig. 5.14 the arrangement of the controls is
different, and here the valves and switches are said to be 'in parallel'. The situation

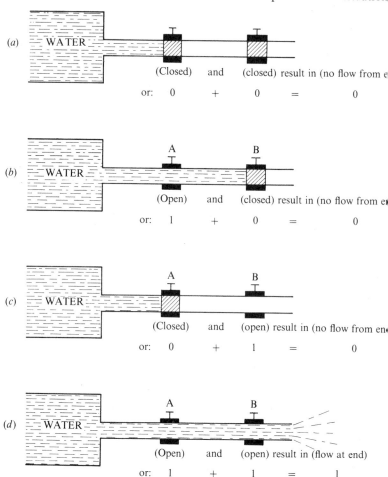

Fig. 5.12

| | Valve *B* | |
| | Open | Closed |
+	1	0
Valve *A* Open 1	1	0
Closed 0	0	0

Fig. 5.13

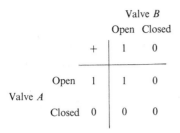

Fig. 5.14

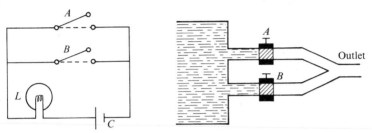

| | Switch *B* | | | Valve *B* | |
| | On | Off | | Open | Closed |
+	1	0		+	1	0
Switch *A* On 1			**Valve *A*** Open 1		
Off 0			Closed 0		

Fig. 5.15

+	1	0		×	E	O
1	1	1		E	E	E
0	1	0		O	E	O

Fig. 5.16   *Fig. 5.17*

is summarised in Fig. 5.15. Before reading on, look at the figure and decide what should go into these tables.
Did you get Fig. 5.16?

The table for O and E, under multiplication, could be written as Fig. 5.5 or could be rearranged as Fig. 5.17. If in the table for switches in parallel we replaced each 1 by E, and the 0 by the O for odd, the tables are seen to have the same shape. Mathematicians are very interested in discovering algebras which have similar operation tables.

+	O	E
O	E	O
E	O	E

+	1	0
1	0	1
0	1	0

Fig. 5.18   *Fig. 5.19*

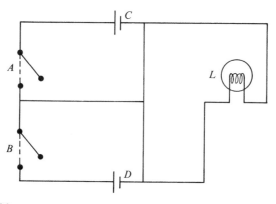

Fig. 5.20

Consider again the table for O and E under addition (Fig. 5.18). Keep the + but replace E by 0 (zero) and O by 1, we have an operation table like Fig. 5.19. Theoretically Fig. 5.19 corresponds to a circuit, but which? One circuit which gives this table is shown in Fig. 5.20, it being assumed that batteries *C* and *D* cancel eac other out if both are on.

The algebra of circuits, which we have discussed in this elementary way is not simply a piece of abstract and amusing notation. By using an elaboration of this it is possible to decide whether complicated electrical circuits are in their simplest form or not, and communications engineers do, in fact, use an algebra of this type in their work.

The algebra of sets

Children know what is meant by a doll's tea-set, a set of spanners or a chemistry set, yet when some parents meet the word 'set' in a mathematical textbook, they suspect that it must have some new and deeper significance. Fortunately 'set' is not given any new meaning.

A set is simply a collection of objects. We must be able to describe the collection in such a clear way that we can decide whether an object is or is not a member of the set. We can do this either by listing the members of the set or by giving a rule which members must satisfy. Some sets are so familiar to us that our language con-

tains special words to describe them: a set of footballers is a team, a set of playing cards is a pack, a set of flowers is a bunch, and there are many more.

Apart from definition, we are often interested in the number of members that the set possesses. Although we must always be able to decide whether a particular object is or is not a member of a certain set, we do not always know how many members it has. The number of apple trees in an orchard at any given time can easily be counted, the number of apple trees in England at any given time is quite definite, although not easily counted. These sets are both *finite* sets. In contrast, consider the set of odd numbers, 1, 3, 5, 7, . . . This is a perfectly well-defined set in that we know whether any given number belongs to it or not, but it goes on for ever, and that is what the dots after the 7 indicate. Such a set is said to have an *infinite* number of members.

In work with sets, we often need to say 'the set whose members are', and to save time we use curly brackets like this: { } . { Types of trees in Windsor Park } is read 'the set whose members are the types of trees in Windsor Park'. { 1, 3, 5 } is read 'the set whose members are 1, 3 and 5'. Often a capital letter is written instead of a set: we might define a set B by writing $B = \{3, 6, 9, 12\}$, and after that refer simply to B.

We often concentrate our attention on a certain background set. For example, we might be talking only about the set of all types of tropical trees or the set of all whole numbers. This background set is often referred to as the *universal* set, and the letters U or $\&$ are used to denote it.

Intersection and union

Working with sets at school level is confined to performing two operations called *intersection* and *union*. Here is an example: suppose on Parents' Evening all the parents are gathered in the Assembly Hall to hear the remarks the Head is about to make; the parents present form the universal set and we would write $\&= \{$ all parer in the Hall$\}$. The Head needs some statistics to support his request to the authoritie for a car park, and he asks the parents to 'put up their right hands if they drove one of the cars used to get to the meeting this evening'. The parents with their right hands raised form the set we will call D, so $D = \{$ parents who drove a car to the meeting$\}$. Those parents who did not actually drive the car will not have a hand up so $\&$ has been split into D and 'the others'; and these others in $\&$ but not in D are denoted by D' and referred to as the set *complementary* to D. Obviously, if we put D and D' together, we get $\&$.

The symbol for 'putting together' or union is $\cup$, and we can write $D \cup D' = \&$. The Head (who enjoys statistics) now asks parents to raise their left hands if they wore seat-belts. If we write $S = \{$ those who wore seat-belts $\}$, then $S' = \{$ those who did not wear seat-belts $\}$, and again $S \cup S' = \&$. A quick glance round the Hall show that some parents have both hands up, and these are those who are members of both sets, D and S: they drove and sensibly wore their seat-belts. If two sets have

'some parents have both hands up'

members in common, the sets are said to intersect, and we refer to this common membership as 'D intersection S', or more shortly as $D \cap S$. The symbol for intersection $\cap$ is sometimes read as 'cap' — it looks rather like a school cap without the peak – so $D \cap S$ is read 'D cap S' and $D \cup S$ is read 'D cup S'. We could make a table like Fig. 5.21. $D \cap S$ would be drivers who wore seat-belts (the ones who would have both hands up) $D' \cap S$ would be non-drivers who wore seat-belts and would have the left hand up; $D \cap S'$ would be drivers who did not wear seat-belts, who would have the right hand up; and $D' \cap S'$ would be those who did not drive and were not wearing seat-belts.

$\cap$	D	D'
S	$D \cap S$	$D' \cap S$
S'	$D \cap S'$	$D' \cap S'$

Fig. 5.21

The empty set

Would it make sense to ask for the members of the set $D \cap D'$? Could there be anyone who drove the car and at the same time did not drive it? $D \cap D'$ cannot possibly have any members. A set with no members is said to be *empty*. In the language of sets we would write $D \cap D' = \{ \ \}$. Inside the curly brackets is just an empty space. Another way of writing an empty set is to use the symbol ⌀. The intersection of any set with its complement is always empty by definition. The empty set seems to fascinate children. They do not always find it easy to understand that there is only one empty set. The set with no members is the set with no members, even though we may previously have been talking about parents, about trees or about letters of the alphabet.

Venn diagrams

Teachers often introduce work on sets by using their classes as the universal set, in much the same way as a set of imaginary parents has been used here. However, the universal set is not necessarily human and able to respond, and we then do what mathematicians always do, draw diagrams to assist our reasoning.

Fig. 5.22 shows a sketch of some boys and girls; together these form the universal set &, and if $G = \{$ girls$\}$, then the set of boys is G'. $G \cup G' = \&$. Some members of & are wearing spectacles; these comprise the set S, and those without spectacles are S'. If we draw a ring round the set G and another ring round the set S, then those in both rings form the set $G \cap S$. Fig. 5.23 shows the essence of Fig. 5.22, without the individual faces. The actual shapes used to represent &, G and S do not matter, but since there are some members in both G and S, these shapes must overlap.

Fig. 5.22

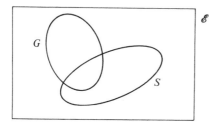

Fig. 5.23

An English mathematician, John Venn (1834–1923), first used diagrams like these to illustrate the relationships between sets, so these are called Venn diagrams. Fig. 5.24 uses Venn diagrams to illustrate the different sets in the table we made when we discussed the sets D (drivers) and S (seat-belt wearers) in the set of parents table (Fig. 5.21). Whatever enclosed region is used for $\&$, it is divided into four regions by the shapes used for D and S, and each of these four regions represents one of the sets in the table. If all the shaded areas were put together, the whole of $\&$ would be covered, so they make up the whole of the universal set.

How could we use a Venn diagram to illustrate that the only parents at the meeting who cultivate an allotment are also car drivers? The allotment holders are part of the set D but none are in D', so the ring for allotment holders must be entirely within the shape used for D. Fig. 5.25 shows this. If $A = \{$ allotment holders $\}$ what is $D \cap A$? $D \cap A$ includes the whole of A, so $D \cap A = A$. If the diagram had been like Fig. 5.26, what would this have meant? Here none of the allotment holders are drivers, and $D \cap A = \{\ \}$ or $\varnothing$.

Now we have said almost enough about sets for you to do a few examples, if you wish. Before you begin, one more point must be made. Suppose you are asked for the members of the set $J = \{$ types of jam in your store cupboard $\}$, and you have four jars of strawberry, five of gooseberry and twenty of red plum. The answer is

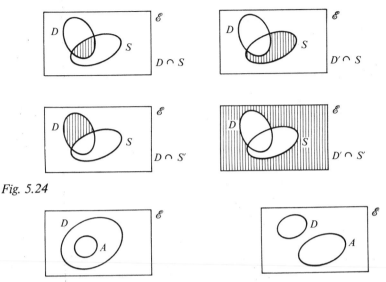

Fig. 5.24

Fig. 5.25   Fig. 5.26

$J = \{$ strawberry, gooseberry, red plum $\}$. If you are asked for the members of
$L = \{$ letters in your surname $\}$, and your surname is Sommers, then $L = \{$ E, O, M,
R, S $\}$. The set L does not indicate how many times any member of the set has been
used. Examples for you to try (you will find the answers at the back of the book,
Answer 11).

(a)   If $A = \{$ letters in the word LONDON $\}$ and $B = \{$ letters in the word
        SOUTHAMPTON $\}$, what is $A \cap B$? What is $A \cup B$?

(b) $R = \{$ even numbers less than 21 $\}$, $F = \{$ 3, 6, 9, 12, 15, 18 $\}$. What is
        $R \cap F$? What is $R \cup F$?

(c) $\mathcal{E} = \{$ types of metal $\}$, $M = \{$ gold, silver, lead, copper $\}$, $L = \{$ silver, gold $\}$.
        What is M'? What is $M' \cap L$?

(d) $\mathcal{E} = \{$ positive whole numbers less than 20 $\}$, $D = \{$ odd numbers less than
        20 $\}$, $E = \{$ 3, 5, 7, 9 $\}$. What is $D' \cap E$? What is $D \cup E'$?

(e)   The Venn diagrams in Fig. 5.27 show 3 possible relations between 2 sets.
        Look back at questions (a)–(d) and decide which diagram you would use
        to illustrate the relationship between the sets (i) A and B, (ii) R and F,
        (iii) $\mathcal{E}$ and M, (iv) M and L, (v) $\mathcal{E}$ and D, (vi) D and E.

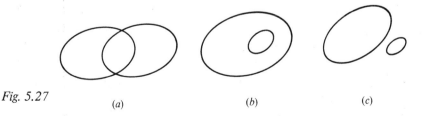

Fig. 5.27
        (a)   (b)   (c)

'of whom 5 are unable to play on Saturdys'

(*f*) A sports club has 40 members, of whom 5 are unable to play on Saturdays. Of those willing to play on Saturdays, 23 play soccer (*S*) and 18 play hockey (*H*). Use a Venn diagram to find out how many are willing to play either on Saturday.

(*g*) A postman had 42 electricity bills and 27 demands for rates to be delivered in a street of 50 houses. What is (i) the smallest, (ii) the largest number of houses that could be getting both bills on the same day?

If you do not have time to do the examples, start again here.

Does it make sense to talk about the intersection of three sets? If we think once more about the parents in the Assembly Hall, we could look round us and see that several were wearing hats. These hat wearers might also have been sitting patiently with both hands up, so they were drivers and had worn seat-belts; they are members of three sets. In Fig. 5.28, the shaded bit in the middle represents the set of hatted, seat-belted drivers. These hatted parents obviously belong also to { ladies }, since no father would sit with his hat on! But to show 4 sets intersecting needs a three-dimensional diagram, and the diagram tends to be more difficult than the idea it seeks to illustrate. There are diagrammatic methods other than Venn diagrams, but we will not consider them here.

Is the intersection of sets commutative? Considering the Venn diagram in Fig. 5.29 shows that it must be: $A \cap B$ is the same as $B \cap A$.

Is the intersection of sets associative? Is $(A \cap B) \cap C$ the same as $A \cap (B \cap C)$? Look at the Venn diagrams in Fig. 5.30 (or, better still, try them for yourself) and you will be convinced that it is.

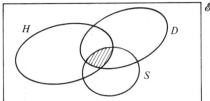

$\mathscr{E} = \{\text{all parents in the Hall}\}$

$D = \left\{\begin{matrix}\text{parents who drove a car}\\ \text{to the meeting}\end{matrix}\right\}$

$S = \{\text{those who wore seat belts}\}$

$H = \{\text{those wearing hats}\}$

Fig. 5.28

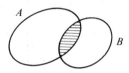

Fig. 5.29

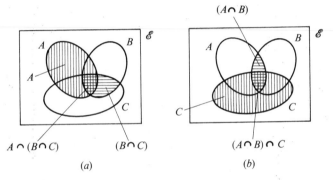

Fig. 5.30

Talking about sets takes a long time. The idea however is easily and quickly understood and children enjoy it. This is not the only reason for teaching it. A lot of logical thinking must go into the handling of sets, definitions must be expressed in careful English, and the universal set must be clearly understood. More important still, the language and the algebra of sets is met in many parts of mathematics, for example, in geometry, simultaneous equations and logic, where they help in clarification and analysis.

So here we have seen three different algebras. What they have in common is 'algebra'. Algebra is at the heart of mathematics and needs a further chapter.

6 Taking algebra further

The algebra of numbers

We want to look more deeply at the ideas involved in algebra. This chapter is slightly tougher than the last, but it starts with something familiar.

Conventional algebra, the sort that most of us were taught at school, remains at the heart of all mathematics. Children brought up in the new way will have had plenty of experience of allowing letters to stand for mathematical abstractions of various types, and will be quite ready to allow letters to stand for numbers.

Solving an equation

The motivation for the algebra of numbers is provided by the desire to be able to solve equations. These often arise in the course of simple problems. A typical example, which might occur in any one of hundreds of traditional elementary texts, is the following: 'Find the number of cows in a field if, when 10 more cows enter, the number present is trebled.'

As a piece of 'real life' the problem is, of course, absurd. But the significant part of solving it lies in formulating and analysing it. The problem could be solved practically by a farmer driving cows backwards and forwards through a gate until he found the answer by trial and error! He would be better advised to sit down with paper and pencil and, by making diagrams, simulate the driving of cows backwards and forwards through a gate. This simulation, that is, using paper, pencils and diagrams in place of real cows in a real field, is known as making a mathematical 'model' of the situation.

We take this process one stage further by using a letter x to stand for the number of cows present in the field. We now represent in symbols the new number of cows when the 10 new ones enter; this is $x + 10$. It has to be equated with three times the original number, that is to say, with $3x$. ($3x$ means $3 \times x$, by convention we miss out the multiplication sign.) We can therefore write an equation

$$x + 10 = 3x.$$

This algebraic equation is a 'model' of the physical situation to be solved. It is easily solved, the answer is $x = 5$. This answer is now referred back to the physical situation and we see that 5 cows meet the conditions. If there were 5 cows in the original field, by adding 10 we should have 15, or treble the original number.

Mathematical models

Another educational aspect of such an example lies in the concept of the mathematical model itself. Although from time to time, particularly in science, occasions will arise when simple equations have to be solved, the mechanics of their solution do not form a very important part of the education of the ordinary man. Certainly the very complicated and unrealistic conglomerations of symbols associated with equation-solving in some traditional textbooks are out of place in a modern course.

Deeper concepts appear if we pursue the idea a little further. Let us suppose that a different problem regarding cows leads to the equation

$$x - 10 = 3x.$$

The solution is still straightforward and we can easily verify that it leads to $x = -5$ since $-5 - 10 = 3 \times (-5)$. This is a perfectly satisfactory algebraic solution. Negative numbers can be just as meaningful as positive ones as anyone with a thermometer in Canada will tell you. However it is quite senseless in terms of this physical situation. Either there is a mistake in setting up the equation or the problem must have been insoluble. There is a condition imposed on the solution set, that is to say, the set of values of x which can satisfy the initial equation. The condition is that x shall be a positive whole number. This is because x denotes the number of cows which can neither be negative nor fractional. It is important to decide (*a*) whether the answer to an equation is the only answer or just one of many, and (*b*) whether the answer makes sense, given its physical meaning.

Logical connectives

The mathematics cannot assist with this last question, but the logic of the solution can be clarified by the use of what are called 'logical connectives'.

The connective that we shall have most occasion to use is read 'implies' and looks like this: ⇒. In most modern textbooks this replaces the device consisting of three

dots in a triangle formation like this ∴ that stands for the word 'therefore'. The use of this connective gives a directional sense to an argument. Here are two statements.

(a)   I am not hungry.

(b)   I have just finished a large lunch.

Now, for most of us it would be true to write (b) ⇒ (a); that is to say, the truth of statement (b) means that statement (a) is also true. It is by no means necessary that (a) ⇒ (b), for 'I am not hungry' may imply 'I am not feeling well' or many other things. This is a directional argument. We say that (a) ⇒ (b) is the *converse* of (b) ⇒ (a).

To consider a piece of mathematical reasoning, it is true that

(a) $x = 5$ and $y = 7$ ⇒   (b) $x + y = 12$

It is not true that (b) $x + y = 12$.   ⇒   (a) $x = 5$ and $y = 7$.

Perhaps x was equal to 2 and y equal to 10, or $x = 3$ and $y = 9$, etc.

Many statements in mathematics are such that the first implies the second *and* the second implies the first. To denote this we use a double-headed arrow like this ⇔; This is read 'implies and is implied by'. For instance: (a) this triangle has two equal sides ⇔ (b) this triangle has two equal angles.

We do not normally letter the statements (a) and (b) as in the previous paragraphs, this has been done to make the statements more distinct.

Here is a step-by-step piece of nonsense.

Suppose $3 = 8$
then $\underline{8 = 3}$
adding $11 = 11$

but this is true, so it is also true that $3 = 8$.

Unfortunately this is not true! To see the error in this argument, it is necessary to insert the connectives.

$3 = 8$
⇔ $\underline{8 = 3}$
⇒ $11 = 11$

If we accept as true that $3 = 8$, this does imply the truth of $11 = 11$, but the truth of $11 = 11$ does not imply the truth of $3 = 8$, the connective is not that way round.

These implication signs are used throughout the mathematics course, but they are particularly important in algebra.

Most of you probably remember something of how to solve simple 'linear' equations of the type given above by adding to, or subtracting from both sides. You probably also remember that there are such things as quadratic equations, and you may remember that these have, or may have, two solutions. Traditional algebra tended to leave one with the feeling that any other sort of equation was automatically insoluble. This, of course, is quite untrue. Methods have been devised to solve all the equations met in science and technology, many of which are extremely complex, to some degree of accuracy.

There are two general methods of dealing with equations which are covered in a modern course. These may help with the most difficult of equations. The first is to draw graphs. The second is to use inspired trial and error, a rough answer leads to a better one, which in turn leads to a better one still. We will have a look at both of these.

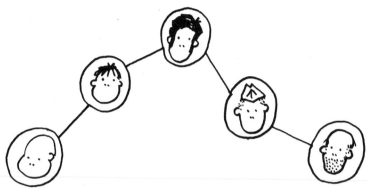

'A graph is a diagrammatic way of illustrating a relation'

Graphs

Let us start by remembering about graphs and how they work. A graph is a diagrammatic way of illustrating a relation. For instance a Venn diagram is, in the broadest sense, a graph. We shall restrict the use of the word, however, to the case in which we are showing the relation between two sets of numbers by plotting points on a pair of axes at right angles. It is most convenient to use graph paper.

Figure 6.1 shows an example of such a graph. It is the heating curve of an electric oven; each point of the paper represents the time in minutes taken to reach a temperature in degrees Centigrade, that is to say, it represents a pair of numbers. Having decided in advance what units the numbers shall be in, we then simply write numbers and enclose them in a pair of brackets. The line refers to one specific oven. Yours will probably have a different curve.

'the heating curve of an electric oven'

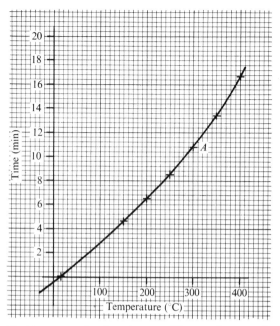

Fig. 6.1. Heating curve for electric oven. Time to heat to given temperature.

For instance, the point A on the curve tells us that it took 10.8 min to reach the temperature of 300 °C. This point is written (300, 10.8). This is called an 'ordered pair of numbers'. The word 'ordered' signifies that the order of the numbers in the bracket matters, that is to say, that this point is not the same as (10.8, 300). The temperature is written first (and is plotted across the page) because it was thought of first; we decided to see how long it took to heat to 100 °C, 200 °C, etc. If we had decided to measure the temperature reached after 2 min, 4 min, etc., the scale and so the order of the pairs, would have been interchanged. From the graph we can easily see how the time varies with the heat required. Although the graph was drawn from a set of only about half a dozen observations, we can use it to see how long it would take to reach any intermediate temperatures, or what temperature it would have reached after any time in the range considered. Read off the time to reach a temperature of (*a*) 220 °C; (*b*) 330 °C; (*c*) 100 °C. Why is (*c*) likely to be the least accurate? Read the temperature after (*d*) 4 minutes (*e*) 15 minutes. (See Answer 12 at the back of the book.)

In this example the figures are obtained as a result of an experiment, but we can also draw graphs from purely theoretical consideration. For instance, we may decide to see the shape of the graph of a relation in which the second number of each pair is twice the first number of a pair. Such pairs are $(1, 2), (3, 6), (1\frac{1}{2}, 3)$, $(-2, -4)$, $(10, 20)$, etc. Obviously, we cannot write down all the possible pairs, and we want some way of summarising all this. If we call the first member of the pair x

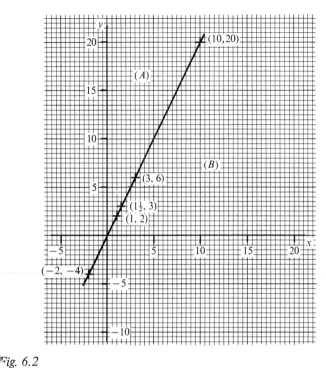

Fig. 6.2

nd the second member of the pair y, we can express the relation as the equation
$= 2x$.

This you will remember is a *convention*. We shall use x for the first member and y for the second member of a pair. Fig. 6.2 shows the points plotted on a pair of axes. They seem to lie on a straight line and it is not hard to prove that they do. Can you see how? If we now worked out further pairs and plotted them, they would also lie on the same straight line. This line is called the graph of the equation $y = 2x$. It is composed of the infinite set of points (x, y) for which $y = 2x$, and we write this:

$$\{(x, y) : y = 2x\} .$$

The colon is read 'such that', and so the sentence in set language reads: 'The set of points (x, y) such that y is equal to twice x'. You will probably have drawn graphs of algebraic functions at school. The only new thing about this is the interpretation of the line as a set of points. Note that the line goes on 'for ever' in both directions, but obviously only part of it can be drawn on a sheet of paper, however large.

Graphs of regions

A straight line drawn on a page defines two sets other than the set of points of which it is composed. There is the set of points on one side of it, and the set of points on the other.

Here are the number pairs of some points on the side labelled (*A*) of the line in Fig. 6.2: (1, 5), (3, 12), (6, 13), $(3\frac{1}{2}, 9)$, (−1, 2), (−1, 8). Here are some points on the side labelled (*B*): $(1, 1\frac{1}{2})$, (3, 4), (6, 1), $(3\frac{1}{2}, −2)$, (−1, −3), (−1, −12). By comparing the first and second numbers in each case, can you suggest relations for the regions (*A*) and (*B*)?

It is not hard to see that for all the points in the region (*A*), the second member is always more than twice the first; this is written $y > 2x$. The symbol > may remind you of a diminuendo sign in music. We help children to remember it by writing BIG NUMBER > small number. We read the symbol 'is greater than'. The same symbol the other way round, <, means 'is less than' and small number < BIG NUMBER. For region (*B*), the second member is always less than twice the first member, and we write this $y < 2x$.

Figure 6.3 shows part of the graphs of

$$\{ (x, y) : y = x \}$$
and $$\{ (x, y) : y = x + 2 \}.$$

Try to say which is (*a*) and which is (*b*).
The paper is divided into three regions

(*C*): $\{ (x, y) : y > x + 2 \}$,
(*D*): $\{ (x, y) : y < x \}$,
(*E*): $\{ (x, y) : y < x + 2 \text{ and } y > x \}$.

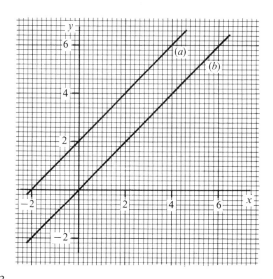

Fig. 6.3

Can you say which is which? Select a point in each, read its number pair and see which of (C), (D) or (E) is true for it. The graphs of these sets are regions. (See Answer 13 at the back of the book.)

Simultaneous equations

In Fig. 6.4 you will see part of the graphs of

$$\{(x, y) : y = x + 2\}$$
and $\{(x, y) : 2y = 5x - 5\}.$

Consider the intersection of these two sets. This is written in set language:

$$\{(x, y) : y = x + 2\} \cap \{(x, y) : 2y = 5x - 5\}.$$

You remember that the sign $\cap$ is the symbol for 'intersection'. The intersection in set language means (luckily!) the same as the intersection in geometry, that is the point marked P. This is the only member that belongs to the sets of points on both lines. It is called the solution set of the equations

$$\left. \begin{array}{l} y = x + 2, \\ 2y = 5x - 5. \end{array} \right\}$$

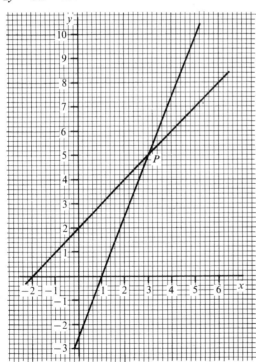

Fig. 6.4

You can see from the graph that the values at this point are $x = 3, y = 5$. The number pair of the point is $(3, 5)$. To test that our eyes do not deceive us, write 3 for x and 5 for y in each equation and see whether the results are true. The first $y = x + 2$ becomes $5 = 3 + 2$, which is certainly true. Make the same *substitution* of 3 for x and 5 for y in the second equation and show that this is also true (see Answer 14). When we are after the point which lies on both, we bracket the equations together and call them *simultaneous equations*.

'There is no need, of course, to use accurate drawing methods'

There is no need, of course, to use accurate drawing methods for two equations such as these. There are simple algebraic methods for solving them which you may remember from your schooldays (see Chapter 12).

Each equation is called a *linear* equation since it will give a straight line when plotted. It can always be recognised by its algebraic 'shape' which is:

$$\boxed{\text{A number}} \times x + \boxed{\text{a number}} \times y = \boxed{\text{a number}},$$

or any rearrangement of this. Algebraically if a, b, c, x, y stand for numbers, then any linear equation has the form

$$ax + by + c = 0,$$

using the convention that a, b, c are known numbers while x and y are unknown ones. For instance, the equation $y = x + 2$ can be rearranged to be $x - y + 2 = 0$, then $a = 1, b = -1$ and $c = 2$.

Inequations

The graphical method, however, is very useful when we need to solve, not linear equations but linear *inequations*. (In the texts these are sometimes called 'orderings' or 'inequalities'.) 'Is greater than' and 'is less than' are relations like 'is equal to'. Instead of $y = x + 2$ we may have $y > x + 2$. In real life relations are more likely to be inequations than equations. For instance, instructions are likely to say 'not more than 3 aspirins' or 'maximum height 10 feet' or 'contents not less than 500 grams'. It is in such cases that we meet inequations.

Look again at Fig. 6.4. Which region represents

$$\{(x, y) : y > x + 2\} \cap \{(x, y) : 2y < 5x - 5\}?$$

You should have indicated the top right-hand sector. What then is the solution set? That is to say, what points of the plane satisfy the two inequations given above? The answer will depend on the nature of x and y.

If x and y are numbers on which there is no restriction, then there is an infinity of possible answers, and all we can say is that they lie in the region bounded by the straight lines in Fig. 6.4, and in the top right-hand sector. Without further information, there is no way of knowing whether some solutions are better or worse than others. If we know that x and y are definitely whole numbers, for example, numbers of vehicles or numbers of men required for a job, then there remains an infinity of answers, but they can only lie at grid points. In Fig. 6.4, some of the grid points in the region indicated are: $(4, 7), (5, 8), (5, 9), (5, 10), (6, 9), (6, 10), (6, 11), (6, 12)$, etc. The inequations become completely soluble if another condition is given; for example, it may be that the sum of x and y should be as small as possible. Then the solution $(4, 7)$ is quickly seen to be the best solution, since for all the other possible solutions the sum of the x and y is clearly greater than $4 + 7 = 11$. In checking this you may have thought that $(3, 5)$ would be better. Can you see why this will not do? (See Answer 15.)

This is the idea behind a topic known as linear programming. This is a rather highbrow term which simply indicates that a problem can be expressed mathematically by a number of inequations. There are various elegant techniques for solving very large numbers of these, but in a classroom we never attempt more than two or three inequations and usually solve them graphically as we have done here.

Non-linear graphs

Graphical solution really justifies itself when we need to solve non-linear equations. Since linear means straight-line, non-linear means curved. Fig. 6.5 shows part of the graph $y = x^3$. Note that there is no need to use set notation when we are not going to use any set theory. It is just as 'correct' to talk about the graph $y = x^3$ as it is to write it out in full as $\{(x, y) : y = x^3\}$. In Fig. 6.5 you will also see the graphs of

(a) $y = 2x + \frac{1}{2}$

and   (b) $y = 2 - x$.

Which one is which? (Test some points found by eye to see which they satisfy; then see Answer 16.)

Now, the line that we have marked (b) cuts the graph of $y = x^3$ at the point

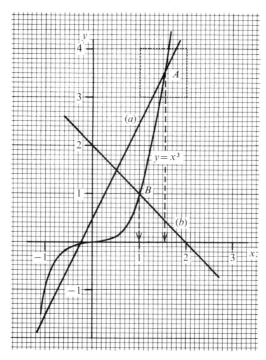

Fig. 6.5

$(1, 1)$. The straight line goes on for ever in both directions, as you would expect, and the graph of the *cubic*, as $y = x^3$ is called, also goes off 'to infinity' in the way shown. There is, therefore, no possibility that there will be any further intersection than the one at B. At the point B, $y = x^3$ and $y = 2 - x$. That is to say $x^3 = 2 - x$ at that point. Drawing these two separate graphs has solved the equation, $x^3 = 2 - x$. The solution is exactly $x = 1$, as we can see by substituting 1 for x, for then $1^3 = 2 - 1$. The shape of the graphs shows that this is the only solution. Why is it so important to substitute and so check graphical solutions? (See Answer 17.)

Turning to the straight line (a), this cuts the graph of $y = x^3$ at three points, of which only two are shown in Fig. 6.5. Where they meet will be solutions to the equation $x^3 = 2x + \frac{1}{2}$. There is a negative solution where x is approximately $-\frac{1}{4}$ and a positive solution at the point labelled A, at which we will look in more detail. It is not very easy to read the value of x at this point, so we have drawn a broken line down to the axes, which shows that the solution is just over 1.5. The accuracy of this solution depends on the skill of the artist, of course, but even more on the scale of the drawing. To get a more accurate solution to this, we take the portion of Fig. 6.5 inside the dotted square and enlarge it four times. This is shown in Fig. 6.6. With this magnification, the graph of $y = x^3$ is indistinguishable from a straight line. The solution can now be seen more accurately. It is about 1.53.

'depends on the skill of the artist'

Now that we have discovered that the positive solution of the equation $x^3 = 2x + \frac{1}{2}$ is roughly 1.53, we substitute to see how accurate this is.

$x^3 = 1.53^3 = 3.58$ (to two decimal places)

$2x + \frac{1}{2} = 2 \times 1.53 + \frac{1}{2} = 3.56$,

quite a close agreement.

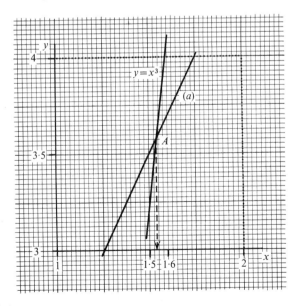

Fig. 6.6

It is now possible to enlarge a small square in the centre of Fig. 6.6 and so get a better solution still. This process of finding roughly where the solution is from a general graph, which may, indeed, be only a sketch, then using a larger scale and getting a better idea of the value of x when the lines cut, and then, perhaps, enlarging again and yet again, is called *iteration*. Compare this with the word reiterate and you will see the sense of it.

Iteration

There is no need whatever for iteration to be tied to graphical solution. It may be purely computational, and this is the second general method of solving equations mentioned above. We should warn you that the next section is a bit harder than any manipulation we have yet met. If you like, skip to the sub-heading 'Working in the algebras', below.

Let us return to the cubic equation whose approximate solution was found graphically above. We write this down and then adjust it by altering the order of the terms and inserting a bracket:

$$x^3 \qquad = 2x + \tfrac{1}{2}$$
$$\Leftrightarrow x^3 - 2x = \tfrac{1}{2}$$
$$\Leftrightarrow x(x^2 - 2) = \tfrac{1}{2}.$$

We start by making a rough sketch and deciding that there should be a root round about $x = 1$. Now, the equation says that $x(x^2 - 2)$ has to have the answer $\tfrac{1}{2}$. If $x = 1$, it follows that $x^2 - 2$ must equal $\tfrac{1}{2}$. This is incompatible with $x = 1$; for if $x = 1, x^2 - 2 = -1$, not $\tfrac{1}{2}$. We can, however, calculate a value of x from this equation; for if

$$x^2 - 2 = \tfrac{1}{2},$$
$$x^2 \quad = \tfrac{1}{2} + 2,$$
$$x^2 \quad = 2\tfrac{1}{2},$$
$$\text{or } x \quad = \pm\sqrt{2.5}.$$

$\sqrt{2.5}$ is approximately 1.6 or -1.6,
but it has to be near $+1$
so it is 1.6 approximately.

Now the fact that these values of x do not agree shows that the first estimate was inaccurate. But it shows more than this, it shows that the first approximation to x, that is, 1, must have been too small. What is more, there is a strong presumption that the true value we are looking for lies somewhere between the two values we have found. Let us take the average of 1 and 1.6, which is 1.3. Now we can start again, and, assuming that x is 1.3, have a look at the value that is required for $x^2 - 2$, and hence work out a new value for this second x. The work is set out as follows (using the symbol $\approx$ to stand for 'is approximately equal to').

Since $x(x^2 - 2) = \frac{1}{2}$,

$$x^2 - 2 = \frac{1}{2} \div x.$$

$$x = 1 \Rightarrow x^2 - 2 = \frac{1}{2} \div 1 = \frac{1}{2}$$
$$\Rightarrow x^2 = 2 + \frac{1}{2} = 2.5$$
$$\Rightarrow x \approx 1.6.$$

The average of 1 and 1.6 is 1.3.

$$x = 1.3 \Rightarrow x^2 - 2 = \frac{1}{2} \div 1.3 \approx 0.38$$
$$\Rightarrow x^2 = 2 + 0.38 = 2.38$$
$$\Rightarrow x \approx 1.54.$$

The average of 1.3 and 1.54 is 1.42.

$$x = 1.42 \Rightarrow x^2 - 2 = \frac{1}{2} \div 1.42 \approx 0.352$$
$$\Rightarrow x^2 = 2 + 0.352 = 2.352$$
$$\Rightarrow x \approx 1.534, \text{ and so on.}$$

This is a slow process but there are ways of speeding it up. The difference between the starting value for x and the final value for x grows smaller each time, and this is of great significance. Using a calculating machine or, better, a computer, the process is easily continued until the difference between the starting and finishing values of x is as small as desired. Once a sufficiently small difference has been obtained, we can say that the equation has been solved to a required degree of accuracy.

Iteration is not a fool-proof process. Sometimes reiterating the procedure can make things worse! There is great educational value, however, in seeing this method applied and knowing that this and the graphical method are applicable, not only to algebraic equations, but also to equations involving trigonometry, etc. A child must not grow up with the idea that there is a small class of equations he can solve and an immensely larger class of utterly insoluble ones. In fact, the computer has made virtually any equation or set of equations soluble to the required degree of accuracy.

Working in the algebras

In this chapter and the last we have had glimpses of three new algebras and a familiar one. In most modern courses there are others as well. They have in common the three essential features: elements, operations and relations; and some time will be spent in a school course on analysing the resemblances and differences between the three.

Some teachers prefer to use an empty square instead of a letter such as x to denote an indefinite or unknown quantity. Let us use this method to make the distinction between an equation and what we shall call an identity. That is, instead of writing $3x + 2 = 8$, we will write

$$3 \times \square + 2 = 8.$$

The pupil has to put a number in the square which will make the statement correct. This highlights the fact that you can mentally fill in the box with any number you like, for example,

$$3 \times \boxed{3} \quad + 2 = 8,$$

$$3 \times \boxed{1000} + 2 = 8,$$

$$3 \times \boxed{-1\tfrac{1}{2}} + 2 = 8,$$

$$3 \times \boxed{2} \quad + 2 = 8,$$

but only the final one of these, of course, is correct. This indicates that 2 is a member of the solution set and we know that this is a linear equation and so it is also its only member. This type of statement is called an *equation*. We contrast it with another type of statement altogether. For instance,

$$3x + 2 = x + 2 + 2x \text{ or}$$
$$3 \times \square + 2 = \square + 2 + 2 \times \square.$$

Now if we fill in the boxes with some numbers chosen at random,

$$3 \times \boxed{3} + 2 = \boxed{3} + 2 + 2 \times \boxed{3} ,$$

$$3 \times \boxed{1000} + 2 = \boxed{1000} + 2 + 2 \times \boxed{1000},$$

$$3 \times \boxed{-1\tfrac{1}{2}} + 2 = \boxed{-1\tfrac{1}{2}} + 2 + 2 \times \boxed{-1\tfrac{1}{2}},$$

and these are all correct. The solution set has an infinite number of members. This type of statement is called an *identity*. In an identity, the right-hand side is often a rearranged form of the left-hand side. There need be no unknowns or blanks in an identity, for example, $3 + 3 = 6$; but if there are xs or boxes, then any member of the universal set can be put in them.

Which type are the following, equations or identities? In each case the universal set has been given.

(*a*) $3 \times 5 + 2 = 17$

(*b*) $x^2 = 4$ {numbers}

(*c*) $(A \cap B) \cap C = A \cap (B \cap C)$ {sets}

(*d*) $\square \cap \mathscr{E} = \square$ {sets}

(*e*) $E + 0 = 0$ {E, 0}

(*f*) $x \times 0 = 0$ {1, 0}

(*g*) $x + 0 = 0$ {1, 0}

In deciding which of the above are equations and which identities, the essential thing to look for is whether each is true just for a few values or for all possible values. (See Answer 18.)

Why is this important? Mathematics teachers have always stated that precision of language is one of their main objects. In mathematics we can only start on a piece of work when we are absolutely clear about our definitions and about our processes. The attempt to formulate a problem in mathematical language is of itself a valuable and frequently difficult discipline. For instance, the answer to the question, 'Where shall we build our new warehouse?' will be the solution to an *optimisation* problem. Many questions will first have to be answered, and so far as possible quantified. For instance:

(*a*) Is the warehouse for storage, distribution, etc.?

(*b*) What types of goods are envisaged, which are the most important? What quantity of each must be stored?

(*c*) Is the warehouse an integral part of a larger operation? If so, is it important that it should be self-supporting?

(*d*) Is capital appreciation important? Is there a maximum cost?

(*e*) Where do the goods come from, go to? How far?

It is poor training for this type of analysis to allow a child to say 'The angles are the same so triangle = equilateral' (using the equals sign to stand for the word 'is'), or to say, 'Let x stand for the cows' (forgetting that x can only stand for a number). Early training in these matters of precision may well count in building up a numerate approach to the problems of later life.

Identity and inverse

There are other things that link these algebras, and indeed all algebras. One is the concept of an identity element, this leads to inverses and we have used these implicitly several times already.

In the algebra of numbers under addition, $3 + 0 = 3$ and $(-8\frac{1}{2}) + 0 = (-8\frac{1}{2})$. The addition of 0 to a number makes no difference to it and 0 is called the *identity element* for addition. What is the identity for multiplication? $9 \times \square = 9$? (See Answer 19.)

Once we have found the identity for an operation such as + or ×, we can sometimes take an element, combine another element with it, and get the identity element for our answer. For instance,

$$3 + (-3) = 0;$$
$$(-8\frac{1}{2}) + 8\frac{1}{2} = 0.$$

The number -3 is said to be the *inverse of 3 under addition* and $8\frac{1}{2}$ is the inverse of $-8\frac{1}{2}$ under addition. When you operate on an element with its inverse, you get the identity. We have used this implicitly in solving equations.

In the equation $2x + 3 = 8$, for example, it is necessary to 'remove' the 3 from the left-hand side. You do this by adding (-3) to both sides, getting $2x = 5$. It is then necessary to 'remove' the 2 from the term $2x$ by multiplying both sides by the multiplication inverse of 2. What is this? You should now be able to solve the equation completely. (See Answer 20.)

Other algebras also may have identity elements. Look again at Fig. 5.1 for the set $\{O, E\}$ under addition. Find a column in the table which is identical with the extreme left-hand column. It is the one under E. This implies that E is the identity element in this algebra. The inverses of the two elements O, E are found by solving the equations $O + \square = E$, and $E + \square = E$. Find the identities and the inverses of the elements in the other algebras given in this section. (See Answer 21.)

Identity and inverse are particularly important because of their connection with the solving of equations. It can be shown that if a and b belong to a set with some rule of combination, and if all equations of the form $ax = b$ are to have single unambiguous solutions which are also members of the set, then

(*a*)　there must be an identity element in that set;

(*b*)　every element of the set must have an inverse;

(*c*)　the set must be associative under this operation;

(*d*)　the set must be closed.

'the set must be closed'

These are the conditions for what is commonly called a *group*.

Relations and functions

We have noted that the 'verbs' of algebra are called relations. In view of what has been said regarding precision of language, this will hardly do for the professionals!

It is in the tradition of modern mathematics teaching, however, to allow a colloquial or partial definition at an early stage, provided it is not incorrect or misleading. A more complete definition then follows when the pupil is ready, sometimes years later. More complete definition and deeper understanding progress hand-in-hand until both are as complete as possible. Little mathematics teaching could ever start if definition and understanding had to be complete from the start. The desire of some university teachers to make the subject too rigorous too soon has been the undoing of some attempts at syllabus revision, particularly in the United States.

A relation links the members of pairs. The members need not be numbers. The relation 'has as his father' links men. For example:

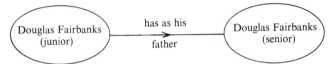

is the statement of a relation in this sense. A relation is not merely concerned with relations in the hereditary sense, however. The relations 'takes place at time' links events with times. For instance:

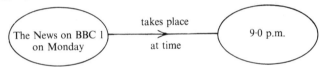

Relations are illustrated by arrow diagrams like these, but when there are many pairs, each set is parcelled up in a ring, as in a Venn diagram. Fig. 6.7 shows part of the same relation, 'takes place at time . . .', with the starting set, $P = \{$ programmes on BBC1 on Monday$\}$, and the finishing set, $T = \{$ times $\}$. As usual, it is enough to

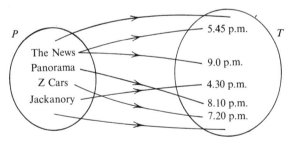

Fig. 6.7

indicate just a few representative pairs so that the idea of the relation is clear. The unattached relation lines indicate that there are more pairs than those shown. The technical terms for starting set and finishing set are *domain* and *range*.

The pairs linked by the arrows are ordered. That is to say:

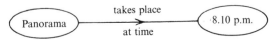

is not the same as

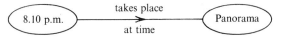

which is obvious enough! There is, however, a relation which runs the reverse way round, from the time to the programme. Try to express it in words. It is called the *inverse* of this *relation*. (See Answer 22.)

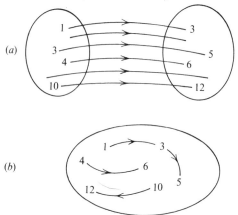

Fig 6.8

Figure 6.8 shows two ways of illustrating a relation linking counting numbers with themselves. The first way (Fig. 6.8(*a*)) makes it clear that the relation links each number with the same number, with 2 added. Verbally, the relation can be expressed 'add 2 gives . . . '. For example:

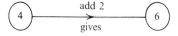

(Putting the relation in words is not always easy without some practice.) The second way (Fig. 6.8(*b*)) makes it clearer that the starting set and the finishing set are the same. The first way is most useful for most purposes. You may notice also that the set of pairs given by this, $(0, 2)$, $(1, 3)$, $(2, 4)$, $(3, 5)$, etc., could equally well be described by the equation $y = x + 2$. There is an equation corresponding to each relation of this type. Children are taught to use either relation or equation with equal confidence.

There is an important mathematical difference between the two relations illustrated in Figs. 6.7 and 6.8. In the relation 'takes place on BBC1 at time . . . ', with the starting set of programmes, there is one programme which has several arrows leading from it. What does this imply? Of course, it implies that this programme (the News) takes place at several different times on a Monday. This is not the case with the relation 'add 2 gives . . . '. Here, each number has just one arrow leaving it,

nd this is where the difference lies. Given the first member of a pair, is there only
ne possible second member? If the answer is 'yes', as in 'add 2 gives . . . ', it is
alled a *function relation*, or simply a *function*. The relation 'takes place at
ime . . . ' is not a function; there are several arrows leading from one of the mem-
ers of its starting set.

The function is a much needed concept in secondary school work and beyond. It
an be made the starting point for manipulative algebra and equation solving. It
eads on to trigonometry and calculus. The concept of relation and function is one
f great generality and power. We have already noted that the sets concerned in a
elation may or may not be numbers. As in the definition of a set, it is not necessary
or a relation even to have a rule that can be formulated. It is enough to give the set
f pairs. Provided each first member is linked with one and only one second mem-
er, it will be a function. For example, $(1, K), (2, M), (54, W)$ is a function. The
tarting and finishing sets have just three members. Fig. 6.9 shows how this would
e illustrated. Such a function is not particularly interesting. In most functions met
y children, there will be rules to define them.

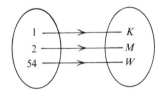

Fig. 6.9

Try to decide which of these relations are functions and which are not.

(a)   {men}------------- has a head of -------------->----------{numbers}
                                size

(b)   {counting / numbers}----------- has as a ----------->----------{counting / numbers}
                                factor

(c)   {counting / numbers}----------- has as its ----------->----------{counting / numbers}
                                largest factor

(d)   {women}------------- has as her -------------->----------{women}
                                daughter

(See Answer 23.)

We think that this should be enough to show how modern courses handle their
algebra. Traditional algebra is written in new ways and involves hitherto more
advanced ideas like iteration. New algebras are developed alongside the algebra of
numbers and are used to emphasise important ideas, commutativity, closure,
dentity and inverse, function.

Despite the changes it is still important for the child to be able to manipulate
algebraic symbols with as much dexterity and accuracy as he is reasonably capable
of. Understanding makes for improved motivation but practice is still required.
What has been omitted is the unnecessary complexity of many of the examples in

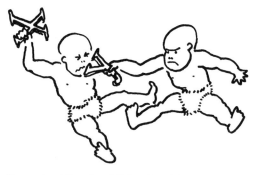

'important for the child to be able to manipulate algebraic symbols'

traditional texts and some of the topics such as advanced factors and algebraic frac
tions which call for skill but make little contribution to understanding of what
algebra means or where it is going.

7 Transformation geometry

How geometry started

The word 'geometria' is a Greek word compounded from gē, the earth, and metron, a measure. Many thousands of years ago the annual flooding of the Nile, brought the fertile silt and water so vital to the economy of the Egyptians, but it also obliterated the field boundaries, and so land measurement was of great interest to the landowners and peasants. Practical men concerned themselves with shapes and sizes of fields while others, perhaps with more leisure, enjoyed the mental experience derived from comparing shapes and finding relationships between their sides and angles. Man's interest in the earth beneath him has always been accompanied by interest in the heaven above, in space and objects which take up space, so relationships were discovered about shapes in both two and three dimensions. It is not

'others . . . enjoyed the mental experience derived from comparing shapes'

surprising therefore that some of the earliest geometrical writings have been found in the eastern Mediterranean.

Learning geometry

In about 300 B.C., Euclid (a Greek), who lived at least part of his life in Alexandria in Egypt, collected together and set down what was then known of geometry. His writings, known as the *Elements of Euclid*, comprised one of the early books to be printed when printing was invented in the fifteenth century A.D. Until very recentl' almost all geometrical teaching in schools was based on Euclid's *Elements*. It is, of course, true that the properties of triangles, quadrilaterals and other geometrical figures are the same now as in 300 B.C., but studies by such men as Piaget on how children learn suggested that methods of teaching could be greatly improved.

Proof

You may remember efforts to learn theorems. Some were to establish, for example, the properties of parallelograms, properties which often seemed too obvious to need proof. Moreover, what was acceptable as a 'proof' depended on what starting points were permitted. Mostly proofs depended on proving triangles congruent by means of sets of rules. In modern courses the properties of geometrical figures are established mainly by dynamic (at first experimental) methods but the emphasis has switched to transformations themselves. Intuition is allowed its place in looking for geometrical relationships – we perceive what we think to be the truth and then set about justifying it. One of the most influential Americans, Professor Polya, says, 'an inductive attitude . . . aims at adapting our beliefs to our experience as efficiently as possible'.

Finding out about geometry

If you would like to try a simple piece of 'dynamic' practical work, fold a rectangular piece of paper once, short edge onto short edge, and cut off a corner of the fold. Open out the corner you have cut off. What shape is it? Remembering how it was made, what must be true about its sides and its angles? What relation does the fold line have to the 'base' of the triangular piece that you should now have, and what does the fold line do to the triangle as a whole? By this simple experiment all the elementary properties of an isosceles triangle can be demonstrated. You will find them in the answers at the back of the book (Answer 24).

Here is another experiment which demonstrates properties of a rhombus. Fold a piece of paper once as before and then edge-to-edge again. Cut obliquely across the corner and open the corner out. What shape will the hole be? What must be true about the sides? Which angles are equal? Are the fold lines the same length? What does each fold line do to the other? At what angle do the fold lines cross? Check with the answers at the back (Answer 25).

For most children such practical demonstrations as these are much more con-
vincing than proofs by congruent triangles! But it is the objectives that are so different. The discoveries are valued only as much as they prepare for new discoveries.
The logic of proof is not a primary objective here.

'Paper tearing can be useful, too.'

Paper tearing can be useful, too. Draw a triangle and cut it out. Tear it into three
as in Fig. 7.1(*a*); now fit the corners together as in Fig. 7.1(*b*). The result certainly

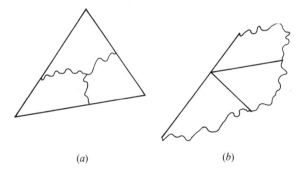

(*a*)   (*b*)

Fig. 7.1

suggests that the sum of the angles of a triangle is two right angles. Now try a four-
sided figure, a quadrilateral. What happens if its four corners are torn off and then
fitted together? Fig. 7.2 shows that a quadrilateral is just two triangles pushed
together, so that the sum of its angles is four right angles.

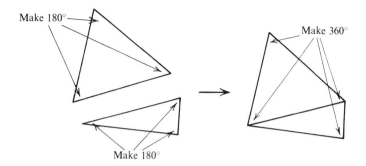

Fig. 7.2

Figure 7.3 shows an insect walking round the sides of a triangle. Through what angle has it turned when it has completed its walk? It has turned through four right angles or 360°. Suppose it walked round the quadrilateral instead, will it have turned through a greater angle? What about a pentagon (5 sides), hexagon, etc.? (See Answer 26.) All these demonstrations can be made logically water-tight if that is what is wanted, but at this stage logical proof is not the main objective.

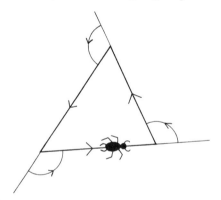

Fig. 7.3

Transformations

Tracing paper is another useful tool in the geometry lesson, especially if the study of the relationship between geometrical figures is approached through what are called *transformations*. Two geometrical figures may be congruent (if cut out they would match exactly), similar (have the same shape but not the same size), or be different from each other. Suppose we have two figures which we think are congruent but which have different positions, if we can define some process by which one figure can be moved without distortion to fit exactly onto the other, we shall know that the two figures are congruent. Movements of any kind are called transformations. This particular type of transformation is called an *isometry* (from 'iso', meaning equal and 'metron', again, a measure. Compare isobar, isotherm and isosceles.)

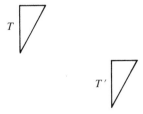

Fig. 7.4

Figure 7.4 shows an object triangle labelled T and its image labelled T'. How could T be moved onto T'? If T is traced, the tracing can be moved onto T' by sliding it over the paper without turning it at all; each point of the triangle will move through exactly the same distance in a straight line and the directions of the movements will be parallel. Such a transformation is called a *translation*.

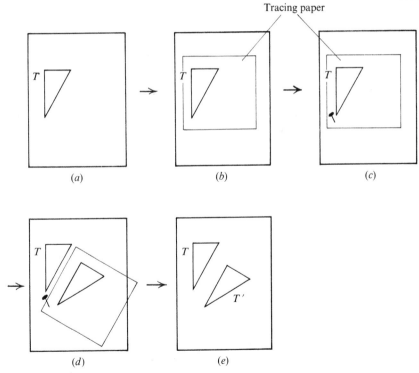

Fig. 7.5

Figure 7.5 shows the triangle T, traced, and turned about a pin into a position labelled T'. Try sliding a tracing of T onto T' without turning. We can obviously not do it. Working backwards, given T and T', there must be some point about which to turn the tracing paper so that T lands on T'. This can be found by trial and error, putting a pin in different positions and turning the tracing; for obvious reasons this transformation is called a *rotation*.

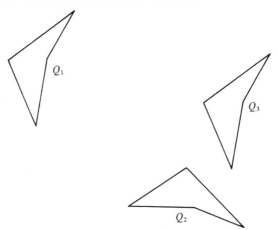

Fig. 7.6

Figure 7.6 shows three positions of a quadrilateral. Try to discover what transformation is necessary to transform (i) Q_1 to Q_2; (ii) Q_2 to Q_1; (iii) Q_1 to Q_3; (iv) Q_2 to Q_3.

You probably discovered that Q_2 can be rotated onto Q_3, but perhaps took a little time to guess where the centre was. Luckily, guesswork can be eliminated and the centre found by construction.

In Fig. 7.7 the point A can be rotated onto the point A' by using a pair of compasses with its point on one of the points O_1, O_2 or O_3 and rotating the pencil in the broken arc which is shown. You will note that O_1, O_2 and O_3 all appear to lie on a straight line. There are, of course, an infinite number of other points O which will do for the centre, such as O_4. The set of all such points do lie on the same straight line and this line is one which passes symmetrically between A and A'.

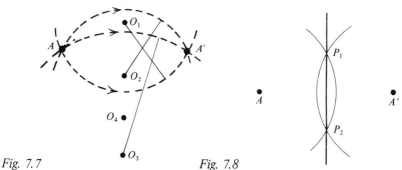

Fig. 7.7   *Fig. 7.8*

In modern parlance it is called the *mediator* of A and A'. Many people prefer the old name of *perpendicular bisector*.

Perpendicular bisector

A perhaps familiar construction which enables you to draw a mediator accurately is shown in Fig. 7.8. The circular arcs centred on A and A' have equal radius and the join of P_1 and P_2, where the arcs meet, is the line of symmetry, that is to say, the mediator.

Now that you can draw a mediator you can find the centre of rotation for Q_2 onto Q_3 in Fig. 7.6. Fig. 7.9 shows the method. First you have to find a centre about which to rotate J onto J'. You do this by drawing the mediator of JJ' which

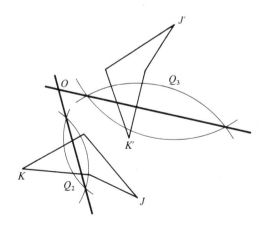

Fig. 7.9

is the set of all possible centres. Then to find a point about which you can rotate K onto K' draw the mediator of these two points also. Any member on the set of points on the JJ' mediator will rotate J onto J', and any member of the set of points on the KK' mediator will rotate K onto K', but there is just one point, the intersection of the two sets or lines, which will rotate J onto J' and K onto K' simultaneously. What is more, once these two pairs have been made to correspond, the remainder of the quadrilateral must fall into place too. Hence the point O is the unique point about which Q_2 can be rotated onto Q_3.

If you want to try one for yourself, trace Q_1 and Q_2. Try to guess the centre of rotation. Then use the trial and error method with some tracing paper. Finally, carry out the construction. You will find that the point lies towards the bottom left of Fig. 7.6.

It is clear that, for this construction for a centre of rotation for one figure onto another to work, the figures must be congruent and 'the same way round'. We will return to this idea of 'ways round' later. The only case that would provide difficulty would be if the mediators turned out to be parallel. It is not hard to see that this would only occur when the figures themselves were in a parallel position like Q_1 and Q_3 in Fig. 7.6. In these circumstances the transformation which we require is a translation, and so no centre need be looked for.

Specification

To specify a rotation requires more than just the position of a centre of rotation. We must also say through what angle the rotation is to take place and whether it is to be clockwise or anti-clockwise. With small children it is usual to start with a turn ($360°$), a half-turn ($180°$), a quarter-turn ($90°$) and so on, and to take the direction as always anti-clockwise. This is the conventional positive direction in the whole of mathematics. Once a point and an angle are given, a rotation is completely specified.

What is necessary to specify a translation completely? (See Answer 27.)

Children can now be challenged to play a game. A drawing is put on the blackboard. It is easiest if a suitable figure is constructed out of hardboard and then traced round. The children of one team then come up and trace another copy of the same figure on the blackboard in a different position. One of the second team then tries to find the transformation. The first team tries to find a position so that it cannot be transformed into the original either by a rotation or a translation. In fact, this is impossible unless the cut-out figure is turned over. Now they begin to see what is meant by 'the same way round' in the previous section.

In technical language we say that the system of rotations and translations in a plane is *closed*. This means that if we start in position 1 with a figure, rotate or translate it into position 2, and then rotate or translate again into position 3, there will always be a single rotation or translation which would take you directly from position 1 to position 3. Compare our previous use of the word in Chapter 5, p. 60.

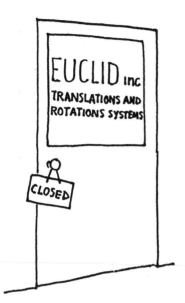

These ideas can be expressed more simply if we set up a notation for transformations. Note how the idea of a set, the intersection of sets, and closure, all previously
met in algebra now appear again, reinforcing understanding and showing how
algebra and geometry are related.

The algebra of transformations

It will come as little surprise to learn that there is an algebra of transformations.
Let us work on squared paper with the usual grid and denote 'rotate 90° anti-
clockwise about the point (2, 2)' by R_1 and 'rotate 90° anti-clockwise about the

'there is an algebra of transformations'

point $(0, 0)$' by R_2. If T denotes the triangle obtained by joining the points $(3, 2)$, $(5, 2)$ and $(5, 3)$, then $R(T)$ is the way we denote the image of T under the rotation R.

Figure 7.10 shows the triangle T and its image after the transformation R_1; this is labelled $R_1(T)$. We now give $R_1(T)$ the rotation R_2; this is written $R_2 R_1(T)$, and the position is also shown in Fig. 7.10.

Similarly, $ABC(Q)$ will mean:

(*a*)   take the figure Q (whatever this is);

(*b*)   give this the transformation C; this transforms Q into $C(Q)$;

(*c*)   then give $C(Q)$ the transformation B; this figure will be labelled $BC(Q)$;

(*d*)   now give $BC(Q)$ the transformation A, giving as its final image in the figure $ABC(Q)$.

It is necessary to stress the order of writing down ABC; can you see why?

Consider $R_2(T)$ and $R_1 R_2(T)$. It is best to make a tracing of Fig. 7.10 and draw the transformations, but a quick check by eye will probably convince you that $R_1 R_2(T)$ is not in the same place as $R_2 R_1(T)$. It will be true for any figure K, transformed in these same ways that $R_1 R_2(K)$ is not the same as $R_2 R_1(K)$. We can write this non-equality like this:

$$R_1 R_2 \neq R_2 R_1,$$

thus stressing that it is a consequence of the transformations R_1 and R_2, not of the particular figure that they are applied to. It is read 'the result of the transformation R_2 followed by the transformation R_1 is not the same as the transformation R_1 followed by the transformation R_2'.

You will note that we write the letters standing for transformations next to each other to denote that one is done after the other. This reminds you that in the

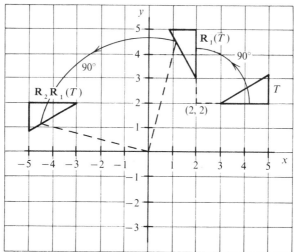

Fig. 7.10

algebra of numbers we write *xy* to denote *x* × *y*. Now 'multiplied by' is an operation in the algebra whose elements are numbers. Similarly, 'followed by' is an operation in the algebra whose elements are transformations. You will have noted that we use capital letters in heavy type to stand for transformations and small letters to stand for numbers.

In Chapter 4 we met the idea of a commutative operation, such as ×, on the numbers (since 3 × 7 = 7 × 3 etc.) and non-commutative operations too. Here we see that the operation 'followed by' is not commutative on the rotations if they are about different centres.

A statement **PQ** = **R** can now be read, where **P**, **Q**, **R** belong to { transformations }. It says that a transformation **Q** followed by the transformation **P** has the same result as the transformation **R** on its own.

'a simple set of rotations of the child himself'

Combining rotations

A good illustration of this (and one which most children find amusing) is obtained by using a simple set of rotations of the child himself through multiples of 90°. We set up the notation as follows:

Q denotes 'rotate through a quarter-turn anti-clockwise about a mark on the floor'.

H denotes 'rotate through a half-turn anti-clockwise about the mark'.

T denotes 'rotate through a three-quarter-turn about the mark'.

I denotes 'rotate through a full turn about the mark'.

The child then stands on the mark and obeys the order '**Q**' by turning to his left, and the order '**H**' by turning left about. He now faces in the same direction as if he

Second transformation

		Q	H	T	I
Followed by					
	Q	H	T	I	Q
First transformation	H	T	I	Q	H
	T	I	Q	H	T
	I	Q	H	T	I

Fig. 7.11

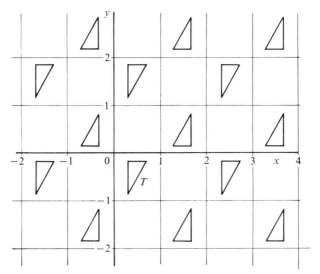

Fig. 7.12

had been given the single order 'T'. He has therefore acted out the relation **HQ = T**. Would **QH** have the same result? (See Answer 28.)

The children can perform a number of these drill movements and note down their result in a table like the one shown in Fig. 7.11. Compare this with the tables in Chapter 5. Why is the 360° turn labelled **I**?

This work with triangles and themselves enables children to explore transformations of the plane using their skill in drawing and their intuition. As they write out their conclusions they will begin to develop notions of proof and completeness. There is the added educational bonus that a new algebra is now being developed, showing an interesting structure similar to those met before.

The infinite sets obtained by successive rotations of the triangle *T* through multiples of 180° about the grid points, that is to say, about (0, 1), (1, 1), (1, 2),

'children . . . will begin to develop notions of proof'

etc., lead to an infinite table and also to the interesting pattern shown in Fig. 7.12. Identify the centres of rotation for some of the pairs of triangles. Combinations of half-turns are either half-turns or translations.

We can now look more carefully at other possible ways of combining rotations and translations. Anyone who has tried to move a heavy wardrobe will know that it is possible to translate by combining rotations; Fig. 7.13 shows the method. We have blocked in one corner so that you can keep track of the wardrobe's orientation (seen in plan). You will see that the rotations do not have to be half-turns. There is a necessary relation between them, however, if the combined transformation is to be a translation. Can you decide what this is? (See Answer 29.) If we continue with our policy of using **R**s for rotations and **T**s for translations, we now see that the following result holds true:

$$\mathbf{R_2 R_1 = T}.$$

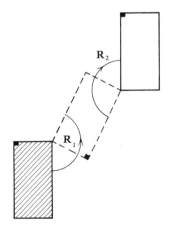

Fig. 7.13

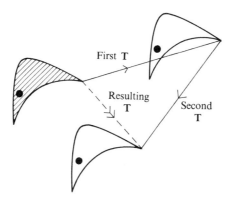

Fig. 7.14

Combining translations

A combination of a translation with another translation is always equivalent to a single translation: see Fig. 7.14. We shall talk about the specification of translations later, when we come to deal with vectors. Algebraically we shall write

$$T_2 T_1 = T_3.$$

A combination of a translation with a rotation gives an image which clearly cannot be parallel with its object. It follows that this must be equivalent to a single rotation, though about a different centre, that is to say

$$R_1 T \text{ or } TR_1 = R_2.$$

We can now draw up a table showing the nature of any possible combination of translations (**T**) or rotations (**R**) in a plane, and this is shown in Fig. 7.15.

'We must now look more carefully at images which are "the other way rour

Opposite isometries

We must now look more carefully at images which are 'the other way round'. In

Second transformation

First transformation	Followed by	T	R
	T	T	R
	R	R	T or R

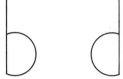

ig. 7.15   *Fig. 7.16*

'ig. 7.16 the letter b is congruent to the letter d but they are clearly not 'the same ⸱ay round'. They are what is called *oppositely congruent*. In the previous sections, ‖l the figures were *directly* congruent, which is the technical language for 'the same ⸱ay round'. Translations and rotations cannot alter the 'way round' of a figure. If ⸱e want to find a transformation, such that b → d as in Fig. 7.16, we must look for ⸱ new one, since neither rotation nor translation will do. It is not hard to think of a ⸱hysical method of transforming the b into the d. We could hold a mirror sym- ⸱etrically between them. In geometry we say that *reflection* is the transformation ⸱ which every point transforms into a point on the perpendicular to the mirror line, ⸱nd the same distance away on the other side. Fig. 7.17 shows a geometrical figure ⸱nd its image (broken line); the construction lines are drawn in lightly. What is the ⸱nage of a point on the mirror line? (See Answer 30.)

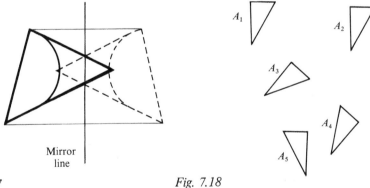

Mirror line

g. 7.17   *Fig. 7.18*

    The next stage in the development is to combine reflections with translations and ⸱otations. This offers further opportunities for developing spatial intuition as well as ⸱ractice in handling the algebra of transformations and, of course, accurate drawing.

    A most interesting investigation takes place when it is asked whether the com- ⸱lete set of isometries in a plane, both direct and opposite, is closed. By now you ⸱ill be used to the technical term and will understand that we are asking if any ⸱ombination of translations, rotations or reflections will be equivalent to a single ⸱ranslation, rotation or reflection.

    Look now at Fig. 7.18. Using **R** to stand for rotation, **T** for translation and **M** for

Transforms into by	A_1	A_2	A_3	A_4	A_5
A_1	I	T	R	?	?
A_2	T	I	R	?	?
A_3	R	R	I	M	?
A_4	?	?	M	I	R
A_5	?	?	?	R	I

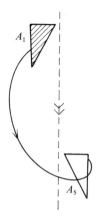

Fig. 7.19   *Fig. 7.20*

reflection (mirror image is another name for reflection, so **M** is used to avoid confusion with the **R** for rotation) and **I** to stand for identity, we can start to fill up a table showing how the triangles which are lettered A_1, A_2, etc., can be transformed one into the other. The table is shown in Fig. 7.19. You will notice that there are a number of question marks. Check this table and try to see why these are so marked.

Investigation will convince you that there are only four distinct ways in which two congruent figures can be placed in relation to one another. These are:

(*a*)   in parallel position and therefore directly congruent;
(*b*)   directly congruent but not in parallel position;
(*c*)   in the mirror position and therefore oppositely congruent;
(*d*)   oppositely congruent but not in the mirror position.

It is the final possibility, (*d*), which causes difficulty (and thus the question marks in Fig. 7.19). We need to define yet another transformation!

The nature of this transformation is most clearly seen by considering $A_1 \rightarrow A_5$ in Fig. 7.18. These are drawn again in Fig. 7.20. The transformation is similar to that experienced by a woodscrew, given a half-turn. Many authors call it a *screw*. In other books it is called a *glide reflection* or simply a *glide*. The figure A_1 is moved forwards in the direction of the broken line, and is also reflected in that line into its new position. It will be specified by giving the axis of the screw, that is, the broken line, and also the distance moved forwards, sometimes called its *throw*.

After a little study it is not difficult to see that a screw will also transform A_2 into A_5. It is harder to see how a screw can transform A_1 into A_4. Indeed, we can only be quite sure that a screw will invariably transform a figure into an oppositely congruent figure, not in the mirror position, if we can produce a construction for finding the axis and for measuring the throw. Look now at Fig. 7.21. X has to transform into X', and by comparing with Fig. 7.20, it is not hard to understand that the middle point, X^*, of XX' must lie on the axis of the screw, and similarly that the middle point, Y^*, of YY' must also lie on the axis of the screw. To find

'Yet another transformation'

the axis, therefore, we construct these middle points and join them up; this produces the broken line which is drawn. The throw is most simply obtained by taking any convenient point, X for example, drawing a perpendicular from X onto the axis, and another from X' onto the axis, and measuring the distance between the feet of the perpendiculars.

What will happen when you apply this construction to two figures in the mirror position? Is it true to say that a reflection is a screw with a throw of length zero? (See Answer 31.)

It is a comfort that we can now see that we can complete Fig. 7.19 by writing **S** (for screw) in place of each of the question marks, and no further transformation needs to be invented to close the set of isometries in a plane. Children find this satisfying and look at wallpaper and lino patterns with new interest, trying to dis-

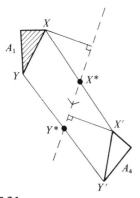

Fig. 7.21

tinguish the transformations which will cause the repetition of all or parts of the design. Try your skill on the design in Fig. 7.22. What transformations can you fin‹

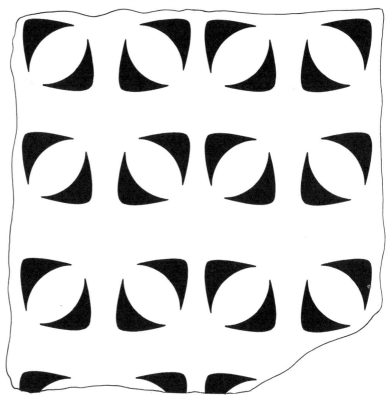

Fig. 7.22. This is part of a pattern that goes on for ever in all directions.

Transformations other than isometries

From congruent figures whose shape and size are the same, it is natural to move on to figures of the same shape but different area, such figures are said to be *similar*. Fig. 7.23 shows a pair of similar triangles; the shape is the same but the size is different. They are drawn with corresponding sides parallel and one is then said to be an *enlargement* of the other. Each side of the large one is three times the length of the corresponding side of the smaller one, or, more briefly, the *scale factor* is 3. If the object triangle (the one we started with) had been the large one, the scale facto‹ of the enlargement would have been $\frac{1}{3}$, since each side of the image triangle (the small one) is $\frac{1}{3}$ the length of the corresponding side of the large one.

What do you find if you join the corresponding corners of the triangles? A quick‹ trial on tracing paper will confirm your guess that the three lines will meet at a point. This is called the *centre of enlargement*. There is an obvious connection with‹

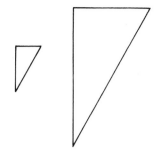

Fig. 7.23

photographic enlargement. How will the distances of a pair of corresponding points from this centre of enlargement be related?

Enough has now been said about the development of transformation geometry to show the general lines and we shall not develop the subject of enlargement.

Where have all the proofs gone to?

In the later stages of an O-level course steps can be taken to write out some of the results touched on in this chapter in a more formal way. Strict formality is not attempted, however, although it is possible to prove all the results of elementary Euclidean geometry from a set of assumptions, or axioms, concerning the nature of transformations.

We hope we have convinced you that there is plenty of challenge in transformation geometry. Its dynamic, experimental basis contrasts with the comparative formality and abstract nature of the algebra. We have also shown you some of the links with algebra of numbers and how the same concepts and the same notation appear in both.

Geometry, however, has not yet exhausted its surprises! In the next chapter we look at a totally different way of exploring space called topology, at a new way of describing geometry called vectors and at the geometry of three dimensions.

8 Some other geometries

Topological transformation

Topological transformations

Transformation geometry is one type of geometry: in the sense in which it is used in school, transformation geometry relates to the sorts of transformation that you met in Chapter 7. In these the image retains a close resemblance to the object figure. In the isometries the figures are congruent, in enlargement, similar. It is good that children should also meet a type of transformation in which this is not the case. This is called a *topological* transformation and leads to a totally different geometry.

The elementary definition of a topological transformation is this. Any part of a figure may be twisted, straightened or stretched at will, so long as it is not broken and so long as two lines are not stuck together.

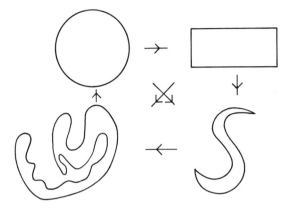

Fig. 8.1

In Fig. 8.1 a circle is shown transformed into a rectangle, then into an S-shape, and then into a non-describable but still joined-up squiggle. The significant thing that they have in common is that they are all closed loops, without loose ends. Physically, figures can be imagined either as drawn on a very stretchy elastic sheet, which is then screwed up or stretched at will, or as line figures made up of very stretchy elastic thread. More of this later. Fig. 8.2 shows some figures, none of which can be topologically transformed into any other. Try to see, in each case, where the attempt to do so would break the rules.

Fig. 8.2

Networks

Figure 8.3 shows four pairs of geometrical figures. Some authors call these *networks*. It is not hard to see that the pairs (*a*) and (*b*) are *equivalent*. This is the word given to a pair of figures when one can be transformed into the other, according to the rules of topological transformations. How can we be sure? One way is to imagine an actual distortion and see how each part of the network can be stretched or shrunk or twisted onto the equivalent one. Another way is to count the number of significant points and the number of line segments joining them. Possibly you did

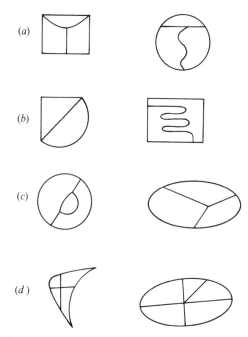

Fig. 8.3

both in checking that the pairs in (*a*) and (*b*) were equivalent. It is useful to have some technical terms here. A point from which lines radiate is called a *node*, the bit of line joining two nodes is called an *arc*. The rules for a topological transformation imply that two networks can be equivalent only if they have the same number of arcs and the same number of significant nodes. The pair of networks in Fig. 8.3(*d*) can quickly be seen not to be equivalent since one has four significant nodes and the other five. Also, the central node in the left-hand network has four radiating arcs whereas the central node in the right-hand network has five. These are called respectively a *4-node* and a *5-node*. We call 4, 5, etc., the *orders* of the nodes. Why are 2-nodes the 'Jokers' of network theory? Your answer to this will explain why we have previously written 'significant' nodes. (See Answer 32 at the back of the book.) From now on we shall ignore 2-nodes and stop using the word significant in this context.

The networks in Fig. 8.3(*c*) are harder to decide about. Both have four nodes and six arcs. You will readily check that each of the nodes is a 3-node. Despite all this, you may have a sneaking feeling that they are not equivalent. To see whether you are right, let us look at the *regions* into which the arcs of each network divide the plane of the paper. We always count the outside of a network as one of the regions. Each network in Fig. 8.3(*c*) divides the plane into four regions; this seems satisfactory. The order of a region, however, is the number of arcs that bound it. Can a topological transformation change the number of arcs bounding a region?

Have another look at this pair of figures and try to decide whether or not they are equivalent. (See Answer 33.)

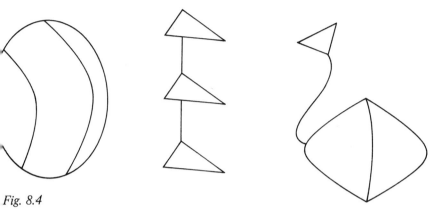

Fig. 8.4

Figure 8.4 shows more networks, each composed of precisely four 3-nodes. Try to decide whether they are equivalent, either to each other or to either of the figures in Fig. 8.3(*c*). (See Answer 34.) It is at this point in a school course that children may meet a *matrix*. This is an array of numbers which can describe the nodes, arcs and regions of a network. We shall meet the matrix in Chapter 12.

Topology at university level is a deep subject with a big literature of great significance for many branches of mathematics. Once past these simple ideas of equivalence the subject rapidly becomes very difficult, which is why there is no subsequent development at school level. The reason for introducing the subject at all, apart from its intrinsic interest and amusement value, is to establish one starting point for the theory of the matrix. As you will see later, the matrix is one of the central concepts of secondary school new mathematics. There is one other reason. The ability to 'see' the equivalence of figures is the 'seeing' that is involved in reading plans or elevations or blueprints, and this is a skill well worth developing.

Vectors

Vector geometry is a tremendously important subject. Vector methods are extensively used in mechanics, electricity, hydrodynamics, etc. It is also another type of geometry and can be used as a starting point for the study of properties of figures, as can transformations. A vector is a line segment of a given length and pointing in a given direction: for instance, it might be a line segment three miles long pointing due north, or a line segment five centimetres long in the same direction as the left-hand edge of this page. Many quantities in physics have a magnitude and a direction associated with them. Velocity, for instance, has a certain magnitude, perhaps 60 kilometres per hour, and a direction associated with it; for example, 60 kilometres per hour along the M1 at a certain point. A velocity of 60 kilometres per hour at

Vectors

right angles to the direction of the M1 at that point would plainly be completely different! Force also has a magnitude and is applied in a specified direction; weight (the pull of the earth on an object) acts directly downwards, and the thrust of a jet engine acts along the axis of its cylindrical casing. Velocity, force and other physical quantities possessing magnitude and direction are called vector quantities. However to be a vector quantity it is not enough just to possess magnitude and direction. The quantities have to obey the vector laws.

A translation, which is a movement by a certain amount in a given direction, is a vector quantity and obeys the laws as we shall see. Children can learn to handle vectors by working with translations. In two dimensions, that is to say, in the plane of the paper, it is easier to give an x-distance and a y-distance than to give a distance and a direction. Fig. 8.5 shows three vectors, all of the same length and in the same direction, and these are said to be *equivalent* to each other. Their specification can equally well be given in two ways:

(*a*)   a distance of 5 units in a direction making $53°8'$ approximately with the x-axis; or

(*b*)   3 units in the x-direction and 4 units in the y-direction. It takes some trigonometry and the theorem of Pythagoras to convince one that these do indeed specify the same vector.

We write the second of these specifications in the form $\begin{pmatrix} 3 \\ 4 \end{pmatrix}$. A column is used to avoid confusion with the point whose co-ordinates are $(3, 4)$. It is important to note that each of the vectors represents the same translation; this is why they are called equivalent. The position of a vector on the graph paper is not significant.

It is obvious that a translation from A to B followed by a translation from B to C is equivalent to a translation from A to C; see Fig. 8.6. We write this **AB** + **BC** = **AC**. Children will not be able to reproduce black type in their exercise books and they

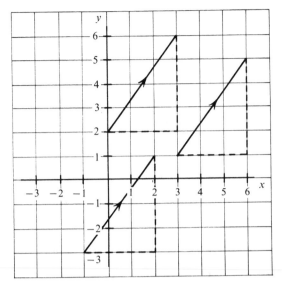

Fig. 8.5

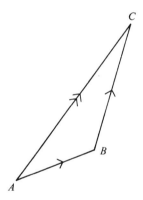

Fig. 8.6

will write either: AB + BC = AC or: $\overline{AB} + \overline{BC} = \overline{AC}$. The distinction between **AB** and *AB* is needed. We take *AB* to mean either the line segment *AB* or the distance *AB*. It is, of course, not true that the distance *AB* plus the distance *BC* is equal to the distance *AC*. Nor is the line segment or bit of line *AB* + the bit *BC* equal to the bit *AC* in any normal sense of the word 'equal'. Note also that, whereas the distance could be referred to equally well as *AB* or *BA*, the vector **AB** is not the same as the vector **BA**. Could these be called inverse? (See Answer 35.)

Vector addition

What is true of the specific translation from point *A* to *B* to point *C*, is also true in the more generalised sense of a vector. A translation whose specification is represented by **AB**, followed by one represented by **BC**, will also be equivalent to one represented by **AC**. Perhaps **AB** may represent a velocity of a current at sea and **BC** may represent the velocity of a ship through this moving water then **AC** will represent the velocity made good by the ship in both magnitude and direction. This is the first and most important of the vector laws.

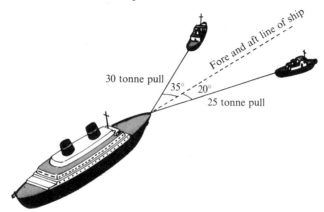

Fig. 8.7

In Fig. 8.7 we have represented a liner being towed by two tugs with forces as shown. The total pull being experienced by the liner can be deduced from the vector diagram shown in Fig. 8.8. Notice that the vectors marked 30 tonne in Fig. 8.8 are both equivalent and, by the vector law of addition, the vector sum is read off by placing the vectors 'head to tail'.

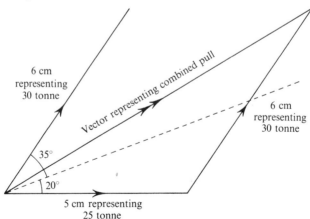

Fig. 8.8

This is the beginning of an algebra of vectors. Let us consider the matter further taking some numerical vectors. Suppose that $\mathbf{AB} = \begin{pmatrix} 2 \\ 1 \end{pmatrix}$ and that $\mathbf{BC} = \begin{pmatrix} 1 \\ 3 \end{pmatrix}$, that is

B is the vector two spaces right and one up, while **BC** is one to the right and three p, then **AB** + **BC** = **AC**. Draw a sketch and so see that **AC** is three to the right and ur up. We write

$$\binom{2}{1} + \binom{1}{3} = \binom{3}{4},$$

e arithmetic of this is easily seen! It is

$$\binom{a}{b} + \binom{c}{d} = \binom{a+c}{b+d}.$$

Iultiplication of a vector

ow for the second rule. **AB** + **AB** + **AB** can simply be written as 3**AB** and this is isily calculated as follows:

$$3\textbf{AB} = 3\binom{2}{1} = \binom{6}{3}; \text{ in general } 3\binom{a}{b} = \binom{3a}{3b}.$$

raw your own diagram. (See Answer 36.)

ector subtraction

AB = $\binom{2}{1}$ it follows that **BA** = $\binom{-2}{-1}$. The identity vector under addition is the ector whose addition makes no difference, and this is plainly $\binom{0}{0}$. It is therefore orrect to say that **AB** and **BA** are inverse under addition. This is useful if we want o find out how to cope with subtraction of vectors. With numbers we can write − 4 = 6 + (−4); that is to say, subtraction is the same as addition of the negative. o with vectors we can write:

$$\binom{3}{4} - \binom{2}{1} = \binom{3}{4} + \binom{-2}{-1} = \binom{1}{3}.$$

teferring yet again to Fig. 8.6 with our numerical specifications for the vectors, we an see that this states that **BC** = **AC** − **AB**, that is

$$\binom{a}{b} - \binom{c}{d} = \binom{a-c}{b-d}.$$

hese rules are enough to be going on with. Later comes multiplication of vectors y other vectors (two types!). Our children will now work on a number of simplifiation problems like these. See whether you can get them right (you can check with answer 37 at the back of the book).

(*a*) Simplify $\binom{2}{1} + \binom{6}{8} + \binom{-3}{-4}.$

(b)   Simplify $5\begin{pmatrix} 3 \\ -3 \end{pmatrix} + 8\begin{pmatrix} 2 \\ -1 \end{pmatrix}$.

(c)   Find x and y if $\begin{pmatrix} x \\ y \end{pmatrix} + \begin{pmatrix} 3 \\ 2 \end{pmatrix} = \begin{pmatrix} 7 \\ 6 \end{pmatrix}$.

(d)   The wind blows at 40 km/h from the north and the bird tries to fly west 20 km/h. What speed and direction does it achieve?

When a good appreciation of how vectors work has been built up, the foundations have also been laid for matrix work for, as we shall see, a vector is in many ways a simple matrix. Furthermore, some of the more important geometrical result concerning centres of gravity can be easily proved by vector methods. These result are needed in statics (a branch of mechanics) and also in statistics. Finally, vectors can easily be extended to three dimensions and even four! Although four dimensions cannot be visualised, it is not hard to build up a set of rules for four dimensions, or even more. So far as this book is concerned, however, we will avoid such flights and stick to three dimensions.

Geometry in three dimensions

One curious thing about traditional teaching of geometry is that it was almost entirely confined to two dimensions. The two dimensions were either on the page of an exercise book or on the plane of a blackboard. The inadequacy of children's three-dimension thinking was strikingly illustrated in an experiment with a class of eight-year-old children at a primary school at Abingdon, Berkshire. They were told the following story:

'You have a friend who is an American Indian; he is intelligent, speaks English, but knows nothing about our methods of measuring lengths. You are going to send him an elephant which you happen to have, and you want him to build a shed big enough for it to live in, and this has to be ready when the elephant arrives. Write him a letter telling him how big to make it.'

The answers were amusing; here are some samples:

'Dear Big Chief Rain-in-the-face,
        You must measure it quite a lot of times across and up.'
'Dear Red-in-the-face,
        Take a piece of banboo and mesure ten arms lengths all round ecept a place for the door.'
'Make a hut roof as tall as the highest tree in the forest.'
'Please make a hut about seven of your strides long and get a piece of wood and measure it 11 strides and then stand it up.'

These were significant, however, as well as amusing. Out of the class of thirty-three children only three thought it necessary to give three measurements to define the size that the hut should be built. We can see at once that it requires length, breadth and height for the hut to be specified. The children found this far from obvious. A teacher at East Stratton, near Winchester, tried a similar question,

egarding a garage for a motor car, on a mixed-age class of seventeen juniors. Only
wo of these gave three measurements. At first we suspected that this lack of
ppreciation of three dimensions was due to their mental development, and that we
night have been hoping for the understanding of a concept which they were too
oung to be able to understand. It was then tried on a grammar school group of
eventeen boys about to take their O-levels. Of these, only three gave three dimen-
ions. In each group, the commonest missing dimension was breadth. The expla-
lation is not easy to give; perhaps the children have photographs in mind involving
ength and height only, but this is less likely to occur in the case of the experiment
with a car. What can be said is that the teaching of geometry must be deficient in
ome way if this is the way the question is answered.

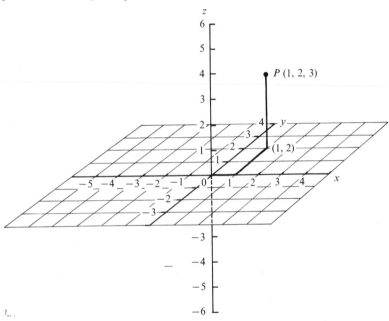

Fig. 8.9

Both transformation geometry and vector geometry can easily be extended to
three dimensions. Fig. 8.9 shows the easiest way to visualise the three dimensions.
A sheet of graph paper with x- and y-axes drawn on it, is held horizontally, and has
a calibrated wire representing the z-axis passing vertically through $(0, 0)$. A point is
now determined by an ordered triple like $(1, 2, 3)$. This is the point marked P. It is
three units above the point $(1, 2)$ on the graph paper. The vector **OP** is $\begin{pmatrix} 1 \\ 2 \\ 3 \end{pmatrix}$. The
laws are the same, but with one extra number in each stock. Try
these:

(a) $\begin{pmatrix} 5 \\ 6 \\ 7 \end{pmatrix} + \begin{pmatrix} 2 \\ 1 \\ 4 \end{pmatrix} + \begin{pmatrix} 0 \\ -1 \\ 0 \end{pmatrix}$;

(b) $5\begin{pmatrix} 6 \\ 7 \\ 8 \end{pmatrix}$;

(c) The glide path of an aircraft lies on $\begin{pmatrix} 10 \\ 8 \\ -1 \end{pmatrix}$. The vector from the aircraft to the runway is $\begin{pmatrix} 1200 \\ 900 \\ -150 \end{pmatrix}$.

Is it on the right path? (See Answer 38.)

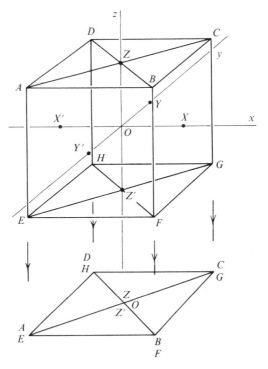

Fig. 8.10

Figure 8.10 shows a cube *ABCDEFGH*, with axis drawn through the centre, so that the positive parts of the *x*-, *y*- and *z*-axes are drawn through *X*, *Y* and *Z*, the centres of the faces shown. If we take the distance *OX* to be one unit, it is easy to see that *X* is $(1, 0, 0)$, *A* is $(-1, -1, 1)$, *G* is $(1, 1, -1)$ etc. We can study the three-dimensional geometry of the cube from a number of points of view.

One is to extend the idea of transformation geometry. The transformation under which *ABCD* → *EFGH* can be a translation defined by the vector $\begin{pmatrix} 0 \\ 0 \\ -2 \end{pmatrix}$.

x-, y- and z-axes

Can you see another possible transformation? The transformation $ABCD \rightarrow HGFE$ is a $180°$ rotation about $X'OX$. Under what transformation would $ABCD \rightarrow FEHG$? (See Answer 39.)

G is the point $(1, 1, -1)$. What will be **OG**? Identify the point B and so write down **OB** and then work out $\frac{1}{2}(\textbf{OG} + \textbf{OB})$. You should find that

the answer is $\begin{pmatrix} 1 \\ 0 \\ 0 \end{pmatrix} = \textbf{OX}$.

(See Answer 40.) But $(1, 0, 0)$ is the point X. Also X is the mid-point of **BG**. Is this an example of a general result? If so, use a similar method to find the co-ordinates of the mid-point of DX.

Yet another advance is to note that the translation of *EFGH* through $\begin{pmatrix} 0 \\ 0 \\ -2 \end{pmatrix}$ will coincide with that of *ABCD* through $\begin{pmatrix} 0 \\ 0 \\ -4 \end{pmatrix}$. The result of this pair of trans-

lations is called a plan view of the cube on the plane $z = -4$. This is a good lead-in to the study of plans, elevations and sections.

So much then for the new syllabuses in geometry. Quite as important, however, as any syllabus is the method by which it is taught. Transformations such as rotation, reflection and translation are dynamic. They invite actual experiment and, when the cardboard and string phase is over, demand imagination. The topological transformations call for special imaginative skill. We are striving towards an ideal where each child rediscovers as much of the main stream of mathematics and travels as far along it as time and his talents allow. Teachers sometimes complain

Rotation	*Reflection*	*Translation*

that their children find this hard. 'The emphasis on discovery is fine', ran a comment from a colleague in the early days, 'but only for those boys and girls who do discover. Unfortunately, not all do.' Yet all children discover something if put in a sufficiently stimulating situation. Sometimes they discover things other than the ones that we hope they will. Sometimes they discover things which are only partly true, or which are untrue in general although they are true in the one special case the child happens to have investigated. The discovery method makes life considerably harder for the teacher, who must be constantly adjusting the situation so that the child may be led in the right direction. Sometimes they hope for too big a step to be made in one. Certainly some children take far longer than others to form the proper concept. The teacher must be patient and wait until the connection has been made. Then, and only then, can the teacher step in and help to rationalise the child's experience, consolidate it and make it available for further development. Geometry provides ample suitable material for this type of development.

If all this takes a long time, it has to be accepted. If school leaving standards, in terms of syllabus covered, are lower, then this, too, has to be accepted. Let us remember that at present the large bulk of secondary children in the United Kingdom fail to reach the present O-level standards in mathematics. Even the more recent CSE examination leaves many behind. Nothing does more damage than pressing on to complete a syllabus before understanding has been reached.

9 Arithmetic

Why teach arithmetic?

It seems hardly necessary to argue the need to be able to do simple arithmetic. The problems of daily life demand little besides addition, multiplication, subtraction, division, a few fractions (halves, thirds, quarters, tenths, together with hundredths or percentages), the area of a rectangle (length multiplied by breadth), and the volume of a rectangular block (length by breadth by height).

No teacher of modern mathematics is going to disagree with learning these, but he may well question the traditional emphasis on some of them, and he will suggest some new and very important topics. These will emerge, first from his desire that arithmetic should fulfil a wider social purpose, that the educated person will by definition be numerate, and secondly from the realisation that ever more powerful calculating aids are coming on the market (see Chapter 11).

As we have already seen, particularly in Chapter 3, modern writers make a good deal of the study of number for its own sake. We shall leave the examples of this already given, number base work, etc., to stand by themselves and in this chapter we shall be dealing with 'bread and butter' arithmetic.

As an example of new method and emphasis we chose a concept which is of great value in a wide variety of everyday situations; this is the concept of *ratio*. As the method of approach we shall use maps and plans. This method has the added advantage that it leads automatically to discussions about accuracy. While there is only one answer to the question, 'How many people are there in the room now?' — for these can be counted — if we ask, 'How far is it to the railway station?' the answer will differ according to the units used, the route taken, the instrument used for measurement, how carefully it is used, and so on.

'How far is it to the railway station?'

Maps and blueprints

We try to find a problem which is real to children, which seems to have some sense in it. We know children react best to questions concerning themselves so for a start, we ask 'How shall we describe to Dad where you sit in class?' It is soon understood that 'next to Judy' is not much help. Most children have seen plans of buildings, blueprints for construction kits, or plans for the construction of cardboard models on the backs of cereal packets and the idea of a picture of the classroom is usually forthcoming. Will it be topological or measured? If the latter, the construction (Fig. 9.1) will involve, in the first place, a detailed discussion of the measurements which will be required. Most of the furniture is happily rectangular so that only length and breadth need be measured. The question which must be answered right at the beginning is 'How accurately have we got to measure?' Here we are instantly confronted with the problems of ratio or scale. The classroom in Fig. 9.1 is drawn on a scale of one centimetre to represent one metre, that is to say, each length is one hundredth of the true size. It follows that a centimetre in the real classroom is represented by one-tenth of a millimetre in the drawing, and since this is very considerably less than the width of a pencil stroke, it is obvious that lengths of, say, 35 centimetres and 36 centimetres in the classroom would be indistinguishable on the plan. This leads to a consideration of what the smallest real distance is that can be represented on the plan. Because of the bluntness of pencils, inaccuracy of rulers etc., one millimetre is as small a unit as a child can deal with. Now, a millimetre on the plan represents 10 centimetres in the actual classroom, and the answer therefore is that there is no point in taking any measurements to a greater accuracy than 10 centimetres. If the width of a table, therefore, is 87 centimetres, we shall register

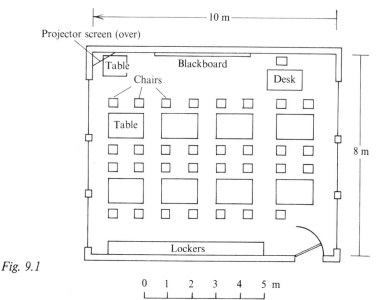

Fig. 9.1

this as 90; if it is 83 centimetres, we shall write it down as 80. All measurements will be corrected to the nearest 10 centimetres.

Learning why and how to work to an appropriate accuracy is educationally most important. So, too, is the ability to correct to an appropriate number of significant figures.

Another question which may occur is, 'How do we find the correct positions for our tables and chairs on the plan?' Discussion usually leads quickly to the idea that we need to know how far it is from the side wall and how far from the front (or back) wall; we need *co-ordinates*. Children sometimes want to mark a co-ordinate grid on the floor of the classroom. In Fig. 9.1 the corner of the teacher's desk, nearest to the blackboard, is 60 centimetres from the wall nearest to it, and 2.8 metres from the nearest wall containing windows (all measurements are, of course, taken to the inner side of the wall). There may be discussion as to whether it is better to measure all distances from the same two walls, or whether it is simpler to take the nearer pair of walls. For practical purposes the latter course is the simpler, although it is not the one which ties in best with the idea of axes in algebraic drawing.

'Once the plan is completed, it should be used.'

Once the plan is completed, it should be used. It cannot be over-emphasised that, particularly in arithmetic, children should see the usefulness of the work that they do. Children may themselves suggest problems; if not, the teacher will be ready with some, like the following:

(*a*)   Is it possible to rearrange the tables so that there is more room for getting around?

(*b*)   If the teacher leaves his desk and visits each pupil once, what is his best route and what distance does he cover? Could this be shortened by a rearrangement?

(*c*)   If an overhead projector is placed on the table near the blackboard, what is the greatest distance of a child from the screen?

We can also use the plan to answer questions about the relative sizes of the furniture in the room. For example, it can be seen that the pupils' tables are approximately $1\frac{1}{2}$ times as long as they are wide, so that the teacher's desk is just a little smaller than one of the pupils' tables, that the windows occupy approximately $\frac{3}{4}$ of the length of the side walls, etc.

At first children may want to continue measuring real tables and real space. It is an important step to see that these questions can be answered from the plan. This connects with understanding that the ratio of corresponding lengths in the plan is the same as the ratio of the corresponding lengths in the actual classroom. It is not an easy step.

This is only one aspect of proportion; and the second aspect is best shown by making a plan of an object which can itself be drawn on the paper. Consider Fig. 9.2. This shows an arrow and a plan of the arrow. Of course, it cannot be said for certain whether this is a large arrow with a half-scale plan or a small arrow with a twice-scale plan, and this ambiguity itself has educational value! Whichever way

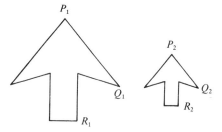

Fig. 9.2

round it is, we can see that for each of the arrows, the slanting sides of the arrow-head are equal, the length of the shaft is twice the width of the shaft, and so on. Next, we compare corresponding lengths in the two plans. For example, the length of the shaft of the larger arrow is twice the length of the shaft of the small arrow. The length of the slanting side of the arrowhead in the larger one is twice the corresponding length in the small one. Indeed, each length in the larger arrow is twice the corresponding length in the smaller one. This is what we mean if we say that the small one is a half-scale plan. Again the concept is not easy, it takes time and built-up experience to grasp fully. We can either compare a length in the large arrow with the equivalent length in the small arrow, or we can compare a pair of lengths in the large arrow with the equivalent pair of lengths in the small arrow. These are two ways of looking at proportion.

Some interesting things can be done with the arrows in Fig. 9.2. The two arrows are similarly oriented; that is to say, they both point straight up in the page. Make a tracing and join up corresponding vertices, P_1 to P_2, Q_1 to Q_2, R_1 to R_2, and so on. What do you find? All these lines, if necessary made longer, should go through a single point. This point is the centre of enlargement which you have met in Chapter 7. Then another question arises; what happens to area when lengths are made twice as great? Although the scale of the smaller arrow is one-half that of the larger one, the area of the smaller arrow is not one-half that of the larger one. Consider just the shaft of the arrows. Trace the shaft onto graph paper, and count the small squares within the boundaries. The larger area is certainly more than twice the smaller area; in fact, the area of the shaft of the larger arrow is four times, not twice, that of the smaller arrow. Can you explain why? (See Answer 41 at the back of the book.)

Of course, this is not 'new' mathematics. Teachers have been teaching much of the detail for a long time. The difference is that the concept of ratio will often be approached, stage by stage, by experimentation and discovery perhaps from plans and blueprints. It is practical experience of ratio which leads to mastery.

Understanding the nature of a problem

So much for building-up an arithmetic concept. Now consider the following problem.

If it is 100 kilometres from Oxford to London, how long will the journey take you at an average speed of 50 kilometres an hour? The answer, 2 hours, comes into the head as soon as the question is read. Now try this one.

A boy can cycle at an average speed of $15\frac{1}{2}$ kilometres an hour, and the distance from his home to school is $2\frac{1}{4}$ kilometres. How long does this journey take? Did you have to think for quite a few moments before you saw how to do it? It is necessary to calculate

$$\frac{2\frac{1}{4} \text{ kilometres}}{15\frac{1}{2} \text{ kilometres per hour}}.$$

Now these two problems are essentially the same. In each we are given a speed and a distance and asked to find the corresponding time. In the first problem the numbers are simple, there is an immediate relation between 100 and 50 and the phrasing almost tells you what to do. The second problem is more tricky, nasty numbers, fractions, the word 'average', and sentences giving facts in a different order. The contrast between them throws light on the nature of mathematical thought. The question can be answered once the pattern is recognised. Mathematics advances, it has been said, by increasing the number of processes which can be performed without really thinking about them. There is a divergence here between the new and the old teaching of mathematics. The traditional view was that this instant recognition was mainly accomplished through drill; that is to say, by working a very large number of the same type of example. 'Don't ask me why, just get on with Examples Set A!'

A modern teacher would say that it is necessary first to build up the proper concept. This may be done in many ways involving experiments and discussion. There will be carefully structured examples. The optimum number differs from child to child. Only once the concept has been grasped will there be some 'reinforcement' drill. Of course, children can forget too. So that regular review of arithmetic skills is needed. There are many techniques for ensuring that this need not be a dull process

The arithmetic of everyday life, as we have noted before, is essentially simple. As teachers, we take advantage of this simplicity to encourage children to think out the steps they have taken to solve a problem. These steps, written down in order (with the help of some mathematical shorthand), produce what is called a *flow diagram*. Flow diagrams can be drawn for every type of mathematical situation. The choice of subject is determined by the stage of development of the pupil, making a cup of tea, starting a car, making a cake, even washing up, are familiar and all contain mathematics. There is some danger for the under-prepared teacher who may find it surprising how complicated some of the things we do with very little thought really are. All flow diagrams have one vital aspect in common: they can only be drawn correctly if the problem is completely understood.

Here is a problem. A hoist has a safe load of 1000 kilograms. It is loaded with 8 crates of soap each weighing 30 kilograms. How many 50-kilogram crates of baked beans can be safely carried in addition?

Before reading further, analyse the solution of this problem for yourself. What arithmetic steps are needed? The constituent parts are as follows:

(*a*)   understanding the English sentences used to describe it;
(*b*)   appreciating the mathematical structure required to solve it;
(*c*)   the calculations;
(*d*)   translating the numerical answer back into physical terms;
(*e*)   a check.

The real difficulty of the problem often lies in stages (*a*) and (*b*), and much of the modern teaching of arithmetic concentrates attention on these very important stages. Drawing a flow diagram concentrates attention on these two stages.

Coming back to the problem, the solution is as follows. We need to find the weight to spare after loading the 8 boxes of soap. This means having to calculate

$$1000 - (30 \times 8) = 760.$$

We must then check that this is in kilograms. We can then divide it by the number of kilograms in the weight of one crate of baked beans:

$$\frac{760}{50} = 15 \text{ (ignoring remainder)}.$$

The answer is, therefore, 15 crates of beans. Carry out the check for yourself by finding the weight of 8 crates of soap and 15 crates of beans. (See Answer 42.)

The flow diagram showing the bare bones of this calculation is given in Fig. 9.3. You will notice that the data needed are described in words, but the actual figures of the problem do not appear. Why do you think that two shapes of box are used in the flow diagram? One reason is connected with the fact that flow diagrams are frequently the first stage in setting up a program for a computer. It is useful to make a distinction between the data of a problem and the calculations. In Fig. 9.3 we have used oval boxes for data and square ones for calculations. It is helpful for children too to be made aware of the logical distinction involved.

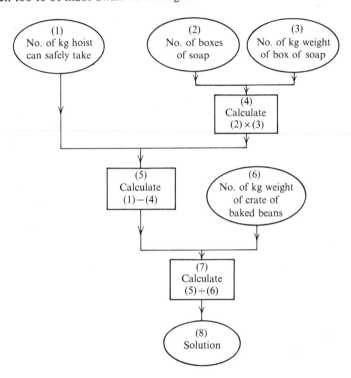

Fig. 9.3

An interesting and testing new problem is to reverse what we have been doing; give a flow diagram and invent a problem to fit it. Fig. 9.4 shows the same flow diagram structure with no words in the data boxes. The same flow diagram would serve for a similar problem in which the numbers of boxes and crates, and perhaps the capacity of the hoist, were all changed. It is also true, however, that the same arithmetic structure will do for many completely different types of problem. Here are two problems: can you discover whether either one has the structure shown in Fig. 9.4? (You can check with Answer 43.)

Problem 1

Find the number of rolls of plain wallpaper required to paper a wall of area 50 square metres, containing eight windows of area $1\frac{1}{2}$ square metres. Each roll of wallpaper contains approximately 5.5 square metres.

Problem 2

A 300-litre cask of sherry is bought by a club for £120, and is distributed to members in 400 bottles, each holding 0.75 of a litre. Find the cost per bottle.

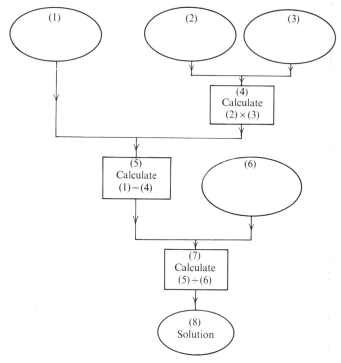

Fig. 9.4

Estimation

The final stage in the working of a problem is to check the answer. This means both to check it against common sense, and to go through and recalculate. The social value of ability to estimate with reasonable accuracy is considerable, and children find the challenge of estimation to be an interesting one. Children like being asked to 'guess the weight of this cake' or to estimate the distance from the school to the station; the time that it takes a boy to read a story; the length of flex required to connect a radio to an extension speaker in the next classroom; the number of tea-spoonfuls in a pound of sugar. There is something quite wrong with the mathemat-

'guess the weight of this cake'

ical education of a child who will give as his answer 'a house 100 feet high', 'a child weighing 400 kilograms' or 'a book with 50 000 words to the page'.

Fig. 9.5

Figure 9.5 shows a straight line. You can find its length by measurement, but try to estimate it without touching the paper. Then ask any friends and relations you can assemble to make their own estimate of its length. When you have obtained as many estimates as possible, calculate the average of the results, that is, add all the estimates together and divide by the number of estimates. This is called the *arithmetic mean* of your estimates. It is probable that this average figure will be close to the true answer, and the larger the number of independent estimates obtained, the more nearly correct the answer will probably be. We return to this in Chapter 10.

We hope we have been able to convince you that traditional arithmetic still has a vital place in mathematics, but that modern techniques of learning can take away much of the tedium. They cannot take away the hard work! Practice is needed from time to time and not every child likes this (though some do).

Sometimes it is said that arithmetic is outmoded now that pocket calculating machines are readily available and seem still (relatively) to be dropping in price. For our views on this, see Chapter 11.

10 Statistics and probability

The last chapter may have left you wondering why lots of estimates are better than just one. You probably found that some estimates of the length of the line in Fig. 9.5 were considerably out. When we assert that the average of the estimates was probably near the length found by measurement, we shall need to consider first what we mean by saying an event is mathematically *probable*.

'what we mean by saying an event is . . . probable'

We shall demonstrate our meaning using a pack of 52 playing cards. You may like to carry out the experiment for yourself. If we cut the pack, the card that is exposed is either a spade, heart, diamond or club. Unless there is something peculiar about the pack, there is no reason to suppose that you are more likely to cut a spade than a heart. Since there are four suits and equal numbers of each, we say that the mathematical probability of cutting a spade is $\frac{1}{4}$. But what does this mean?

We cut the pack 8 times and recorded the suit turned up each time. The result was ♥♥♠♣♦♣♠♥. Out of the 8 cuts, spades came up twice, and 2 is $\frac{1}{4}$ of 8. Is this what we meant by saying that the probability was $\frac{1}{4}$? No, for hearts turned up not twice but three times; how shall we account for that? When you think about it, it is

'Unless there is something peculiar about the pack'

not in the least unlikely (in the homely sense of the word) that hearts should come up 3 times out of 8 cuts; even 4 hearts out of 8 would not cause much comment. However, 8 hearts out of 8 cuts would.

We tried cutting the pack 40 times, recording the number of hearts; then 200 times; then 1000 times. Here is a table showing what happened.

Number of cuts	Number of hearts turned up	Number of hearts as a fraction of total cuts
8	3	$\frac{3}{8} = 0.375$
40	12	$\frac{12}{40} = 0.3$
200	47	$\frac{47}{200} = 0.235$
1000	246	$\frac{246}{1000} = 0.246$

The proportion approached nearer and nearer to the theoretical proportion of $\frac{1}{4}$ or 0.25. This helps us to understand more clearly what we mean by saying that an event has a probability of $\frac{1}{4}$ of occurring. As we make more and more cuts, the proportion of hearts gets nearer and nearer to $\frac{1}{4}$. So also does the number of clubs, diamonds and spades cut, of course.

Now there is another way of looking at this. We carried out a different series of experiments. We started with 8 cuts of the pack once again. We got ♥♦♥♣♠♠♦♣, and noted that 2 hearts came up. Then we cut 8 times again and got ♦♦♥♠♦♣♣♠. This time there was only 1 heart. We went on, noting the number of hearts each time, until we had done it 80 times. We set the results out in the following table.

0 hearts came up in 7 of the experiments
1 heart came up in 21 of the experiments
2 hearts came up in 26 of the experiments
3 hearts came up in 17 of the experiments
4 hearts came up in 8 of the experiments
5 hearts came up in 1 of the experiments
6 hearts came up in 0 of the experiments
7 hearts came up in 0 of the experiments
8 hearts came up in 0 of the experiments
TOTAL 80

You can see from this that 2 hearts came up most frequently, 26 times in the 80 experiments, or on 32% of the occasions. However, 1 heart also came up almost as frequently, with 3 hearts not far behind. Remember that 2 hearts is the theoretical number. So it is not surprising that 2 hearts and its nearest alternatives, 1 heart and 3 hearts, between them came up 64 times, that is, on 80% of the occasions. This is an illustration of what is called *normal* behaviour. This is a statistical term, but we use the word 'normal' because under the right circumstances this sort of pattern is what 'normally' occurs.

These circumstances will occur whenever there is a most likely event, and when departures from this are just as likely to be over as to be under. For instance, in an examination the marks will probably be clustered around an average mark and very high and very low marks will be correspondingly rare. It is the same with people's heights. Most people are of or near average height, with few extremes.

If in a distribution one height occurs more frequently than any other, this is called the modal height or *mode*, and if we lined a lot of boys up and found the middle boy, his height would be called the *median*. If there were an even number of boys, the two middle boys' heights would be added and divided by two, and this would then be the median height. The arithmetic mean, the mode and the median are all representative of a collection of heights. They are a sort of middle, though they are not usually equal to each other. They are all types of average, though in ordinary speech 'the average' is usually taken to be the *arithmetic mean*.

It is very important to note that an individual figure or measurement is much more likely than not to differ from the average, whichever one is chosen. The extent of this variation can, however, often be predicted, and a variation which is very much greater than that expected is sometimes a sign of some significant trend.

A good example of the application of this theory is to consider the statistics for road deaths. Here are the figures for the numbers killed during the five-day Easter periods in nine particular years.

1960	1961	1962	1963	1964	1965	1966	1967	1968
104	95	83	113	86	128	121	121	82

You may like to work out the arithmetic mean, median and mode. (See Answer 44 at the back of the book.)

A general tendency from year to year is not apparent. Variations from the arithmetic mean occur from year to year as we expect. These depend on many factors, such as the date of Easter and the weather. Before any elation can be felt for a decrease, a downward trend must be detected or the amount of the decrease must be compared with the level of variation that is to be expected from natural variations and shown to be 'significantly' more, thus indicating perhaps that new safety measures worked. The calculation of this is beyond the scope of the work that we will be doing at this time, but it is important that children have their attention drawn to the fact that every time a figure drops below average they should not start to cheer, nor to feel depressed when it goes above average. This is 'normal' behaviour.

The misuse of statistics

Most accidents occur at quite short distances from the car-drivers' own homes. This does not mean that the motorists are particularly careless or irresponsible in their own localities. It simply means that many more short journeys are made than long ones. More journeys mean more opportunities for accidents.

Crimes of violence are increasing in number; so are the sales of pop-records; so also are the sales of dishwashers. It is as illogical a use of the figures to associate the increase of violence with the increase in the sales of pop-records as it is with the increase in sales of dishwashers. All three trends are perhaps reflections of another trend altogether, namely increasing affluence.

For purposes of advertisement, statistics are more often displayed by means of diagrams than as sets of figures. Most children will have been introduced to graphs, bar charts, pie charts and pictograms before they reach their secondary school, and as this topic is developed, an important aspect of the work is a critical discussion of the use and misuse to which each method can be put. The diagram in Fig. 10.1 has a fault: can you spot it? (Check with Answer 45.)

While discussion of statistics is taking place, it is quite a salutary experience for older pupils to try to frame a questionnaire on some topic on which statistics are required: it is surprisingly difficult to frame a question in such a way that it means the same to everyone. If it means something slightly different to two people, what reliance can be put on their answers? Care must always be taken when trying to draw conclusions from statistical data.

It is not statistics that are dangerous, only the ill-informed use of statistics, and few will disagree that children need to be taught the difference between valid deductions from a set of figures, and invalid deductions.

You may be asking where the calculations are that one expects with statistics, the averages, the frequency diagrams and so on. These are important and a modern course will give plenty of opportunity for collecting, presenting and averaging information. What matters most, though, is how this information is to be used. In this chapter we have shown just a little of how children are taught to discriminate and how to be sceptical.

Growth of exports since 1958

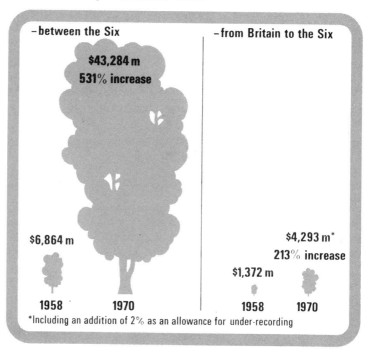

Fig. 10.1. Source: Britain and Europe, 1971. Reproduced by permission of Central Office of Information.

11 Computation and computing aids

Some children enjoy adding, subtracting, dividing, multiplying, etc. Probably most get satisfaction from getting their answers right. Unfortunately there are many who, for various reasons, find accurate computation very difficult. How important are correct answers?

For the abler child it is sometimes said that the answers matter little, it is the method that counts. Certainly examination marking is arranged so that slips in accuracy lose comparatively few marks. If it is understanding that is being measured, a slip or a misreading is not significant compared with a child's overall grasp of the question asked. When we come to real life, however, correctness becomes essential. Books will not balance, wheels will not turn, bridges may fall down if the answers to preliminary calculations are incorrect. There has been controversy over the relative importance of these points of view in the education of children in school.

For the less able it is obviously very depressing to get a succession of incorrect answers. This can lead to correct methods being rejected as incorrect, or unhelpful, or to a blockage.

In problem-solving situations it makes sense for both groups of children to use whatever mechanical aids are available, so that correct answers can be obtained as quickly and easily as possible. Men have felt this since early times. The counting frame, or abacus, has been used as an aid to calculation for over 2000 years. Mention was made in Chapter 3 of the spike abacus, used to improve understanding of place value and arithmetic in differing bases. Fig. 11.1 shows a photograph of a modern calculating abacus. Sophisticated forms of this are used, particularly in the

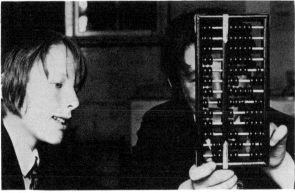

Fig. 11.1

East, for complex work and skilled operators can produce the answers at astonishin
speed.

In this chapter we shall look at four aids which you will find in the modern classroom and will finish by considering the role of the electronic computer itself.

Logarithmic tables

The table of logarithms has been the traditional calculating aid in the secondary
school for many years. 'Taking logs' enables the processes of multiplying and
dividing to be replaced by the simpler processes of adding and subtracting respectively. You may be familiar with this.

Let us suppose that we want to multiply 2 by 3, using logs. The process is as
follows. We describe it on the left and set down the results, in conventional form,
alongside, on the right.

		Number		*Logarithm*	
Look up the log of 2					
in the tables		2	→	0.3010	
Look up the log of 3					
in the tables	X	3	→	0.4771	+
Add the logarithms together				0.7781	
Look up the antilog of 0.7781					
in the tables		6	←		

For the 'antilog' you are effectually using log tables backwards, you are looking for
the number whose logarithm is 0.7781, using an inverse process, in fact. If you
would like to know more about why logarithms work, you will find a fuller treatment in many school courses, for instance in *SMP Book 4*.

For many years it was the custom to teach logarithmic calculation from about
age 13 onwards, and traditional O-level examiners have set much store by ability to
handle quite complicated calculations by this method. In fact it has been hard to
find many people, outside the schools, who use logarithms to any extent. Most
modern courses have therefore stopped insisting on them. Do not confuse this,
though, with the theory of logarithms which remains a very important study at
A-level and later stages.

The slide rule

The most favoured replacement for log tables has been the slide rule. The days when
a sound model could be bought for five shillings are, alas, past, but they are still
relatively cheap. The smaller models slip into the pocket, and learning to use them
brings a bonus which will be referred to later.

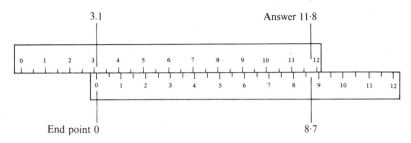

Fig. 11.2

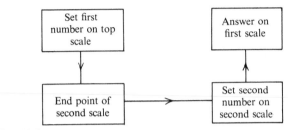

Fig. 11.3

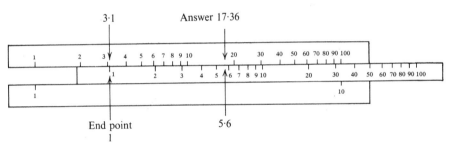

Fig. 11.4

Figure 11.2 shows two rulers arranged so that they will produce an approximate answer to the addition sum   3.1 + 8.7 = 11.8.

Put together like this the two scales make an addition slide rule. To underline the method, Fig. 11.3 shows the flow diagram. As always, this does not contain the numbers to be added, only the instructions to be followed. Addition slide rules do not exist, so far as we are aware, since addition is too easy a process to require such approximate methods.

When we talk of 'a slide rule' we mean a *multiplication* slide rule, and one is shown in Fig. 11.4. It is set to produce an approximate answer to the multiplication sum   3.1 × 5.6 = 17.36.

Why have we taken trouble to say 'approximate' in both this section and the preceding one?

It is significant that the flow diagram for this is the same as that which describes the operation of the addition slide rule, namely that in Fig. 11.3. The secret of the slide rule is that the distances from the endpoint marked 1 to the numbers 2, 3, 4, etc. are proportional to the logarithms of these numbers. That is why the interval between 2 and 3 is greater than that between 3 and 4; the intervals get smaller as we go from left to right along the rule. The slide rule actually adds logarithms. As we have seen, this results in multiplication. If this, very condensed, explanation confuses you, do not worry, it is not necessary to have mastered this to be able to read on.

Using a slide rule quickly and accurately is not easy. Physical dexterity is needed to position the sliding scale exactly where needed, the scales are not too easy to read and it is frequently necessary to infer the size of a number intermediate between two numbers marked on the scale (this is called *interpolation*). Of course maximum accuracy of setting and reading is essential, if correct answers are to be obtained. A further complication is that the same setting is used to compute, say 0.31 X 56 000, as to compute 3.1 X 0.56; in each case the position of the decimal point in the answer has to be found by an independent operation with which the slide rule will not help you.

Perhaps we have put you off slide rules! In fact this instrument is in daily use in many technical departments, laboratories, offices etc.; it is part of the real world of computing. Once mastered, it is quick to use and easily transported. The bonus comes from one of its defects. Because it is necessary to place the decimal point by inspection, it is necessary to think hard what one is doing. Thinking is an important part of education! The by-product is skill in approximating, in getting a feel for the expected size of an answer, for scale, for relations between data and solutions.

Barrel-type calculating machines

A typical machine which may be found in both primary and secondary schools is shown in Fig. 11.5. It will carry out all the four rules (not just multiply and divide, as the slide rule). A large machine will work up to a dozen significant figures (the larger slide rule will give four). The great advantage for small children is that the machine is set by levers to 1 or 2 or 3 etc., so that there is no interpolation. It is what we call a *digital* machine, it works in whole numbers.

It has been found that slower children, who find calculation difficult to the point of slowing their progress, can learn to deal with quite complicated problems if they gain confidence in the machine, that it will always give them correct answers. It seems that learning takes place more easily if anxiety about accuracy is removed. When such machines have been used in primary schools, it has been noticed that work on a machine stimulates pencil and paper calculations. Children like to check up on it!

At one time it was thought that this type of calculating machine would be more and more widely adopted in schools and that they might work a minor revolution

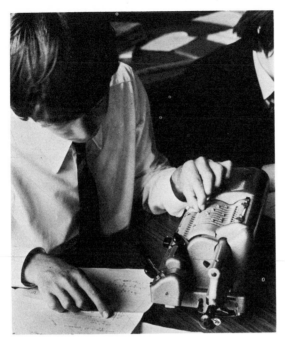

Fig. 11.5

on their own. In fact they are rapidly being superseded by the type of calculator we mention next.

Pocket electronic calculators

It is no exaggeration to say that we are witnessing a calculating revolution through the medium of these little gadgets. As the price drops, in real terms, not merely relatively, millions are being sold to adults and to children. This is being helped by a high-pressure advertising campaign. Their performance far surpasses that of the barrel machines, for they are easier to set, read and check, they are silent, slip into the pocket and many of them are capable of a wider range of automatic calculations. Their advantage over the slide rule is even greater, they have far greater accuracy, and, like the barrel machines, are digital.

If you have such a calculator before you, there is no need for pencil and paper calculations. You can forget fractions and decimals and quite a bit about proportion and ratio. Up to now the implications of everyone having one have scarcely been thought out. If we could be sure that everyone had ready access to one at all times, there would be a real query over the need for memorisation of tables, practice in arithmetic and the rest. The emphasis could swing even further towards approximations and the appropriate checks, perhaps by de-calculating, that is by working

from the answer backwards to the number started with (inverses in fact). This swin
has already started with the wider use of the slide rule, but little pocket calculators
can be operated by children who would not have anything like the ability needed t
master the slide rule.

In practice, teachers will continue to teach arithmetic skills for many years yet
and we are seeing something of a reaction against even the present relaxation of
emphasis on computational accuracy by traditional methods.

There are further implications. Widespread introduction of pocket calculators
into examination rooms would necessitate radical changes in the setting of examination papers. There would be little point in setting calculation questions if they
could be done at the touch of a button or two. On the other hand, problems could
involve more realistic numbers. At present there have to be 500 children in a school
100 heights to average, percentages have to be 5% or 10%, just in order to avoid
tedious working. It is as easy to set $11\frac{1}{4}\%$ as 10% on a calculator.

A final advantage of these machines is that they enable quite young children to
study numbers for their own sake. They might do this by working with very large
numbers, normally too daunting, by pursuing sequences up into the millions, or by
using the special facilities of some machines to build up number patterns like that i
Fig. 1.2.

Computers

There is no doubt in our minds that the most significant invention of the twentieth
century so far is the electronic computer, and we say this being fully aware of the
claims of television, penicillin, satellites, nuclear power etc. The computer can store
and process information at such a speed that there is almost no longer such a thing
as a mathematical problem that is insoluble. This makes mathematics available to
control and predict the most intricate, complex and huge-scale operations. Whether
this is to the ultimate benefit of mankind remains to be seen. It will depend on the
vision and ability of children now in school. They must grow up with a working
knowledge of this creation of man's mind, what it can do and, perhaps more import
antly, what it cannot do. They must know enough to react, whether as voters, tax-
payers or managers, if computer-power seems to be threatening human dignity,
privacy and even liberty. It is not stating the case too strongly to say that there is
real danger of this being so.

However, it is as a teaching aid, perhaps one should say, as a teacher, that the big
computer has its greatest potential for schools. When properly programmed, it is
possible to talk to such a machine. You do this by typing messages to it or pushing
buttons. It will not be long before the human voice will be acceptable. The computer provides each child with his own indefatigable teacher, who knows everything
never makes a mistake, can show pictures, draw diagrams, play tapes. What the
psychological effect of a whole school career face to face with one would be, we ca
not imagine.

'it is possible to talk to such a machine'

But this is still some time away. At present, few schools have links with really big machines, though more have 'desk-top' computers (Fig. 11.6) which have some

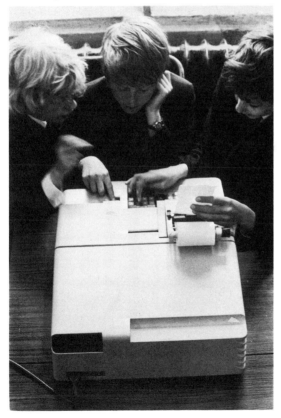

Fig. 11.6

of the big machine's capabilities, or have access to commercial machines through mailing punched instruction cards.

Besides sociological aspects, there is mathematical education in learning about computers. They are instructed by means of programs (spelled like that). To write a successful program requires skill and accuracy and understanding of a particular machine's logic. The steps are as follows:

(*a*)   define the problem precisely;

(*b*)   draw up a flow diagram showing how the computing will proceed, what will come before and after what, and so on;

(*c*)   translate these steps into the proper language;

(*d*)   run a test program to check that there are no errors.

These skills are analagous to mathematical ones. Although computer programmers are certainly not all trained in mathematics, programming is a very useful topic in modern syllabuses. They can start with the most simple ones at age 12 or 13 and provide meaty problems up to school leaving age and far beyond.

Machines, of whatever size, have not yet taken over the task of organising calculations, of posing the questions or of analysing and acting on the results. In this chapter we have attempted to look a little way ahead. For the moment the place of arithmetic is secure!

12 Matrices

Because of the need to keep the material in this book within manageable limits, the previous chapters on secondary school mathematics have mentioned only a few important topics and kept to traditional subject boundaries. Now we want to give you a glimpse of new school mathematics with which you are probably not familiar. This is just a special way of storing information; it is called a *matrix* (plural matrices) and has already been mentioned in Chapter 8. You may think that information-storing does not sound particularly exciting. However, it turns out that matrices combine in interesting and useful ways with each other — thus providing yet another algebra — and that they have applications also in arithmetic, geometry, trigonometry and much besides, and provide a link between these apparently different branches of our subject. Matrices are therefore worth a chapter to themselves.

There are a number of ways of introducing the idea of a matrix. One way is through topology and networks. We are going to start with some arithmetic, and see how to use matrices to organise calculations.

Purchase and price matrices

Suppose a small hotel has deliveries of bread and rolls only on Monday, Wednesday and Friday each week. On Monday in one particular week the order is for 10 large white loaves, 4 small brown loaves and 30 rolls. On Wednesday the order is for 8 large white loaves, 3 small brown loaves and no rolls. On Friday the order is for 12 large white, 5 small brown and 40 rolls.

Notice how confusing it is to present the information in 'essay' form. It is much more helpful to set it out in a table, like this.

Day	Large white loaves	Small brown loaves	Rolls
Monday	10	4	30
Wednesday	8	3	0
Friday	12	5	40

Having noted that the rows refer to days of the week (Monday, Wednesday and Friday) and the columns to types of loaf (large white, small brown, rolls) we could now omit the headings and write this information even more briefly:

$$\begin{pmatrix} 10 & 4 & 30 \\ 8 & 3 & 0 \\ 12 & 5 & 40 \end{pmatrix}.$$

This array of information, without headings or units, enclosed in brackets is a matrix. It is often useful to be able to identify a particular matrix, or talk about a matrix in general terms, by using a capital letter, for instance the letter A. Books usually use heavy type like this: **A**. Children usually write it with a twiddle under it like this: A̰. What does this remind you of? Since we shall make more use of the matrix above, we shall write:

$$\mathbf{A} = \begin{pmatrix} 10 & 4 & 30 \\ 8 & 3 & 0 \\ 12 & 5 & 40 \end{pmatrix}.$$

Now let us get on with the story of the hotel's orders for the baker. We suppose that the weekly requirement remains the same for five weeks, but then has to be changed. The new matrix, we will call it **N**, which gives the quantities of the three types on the three days of the sixth week is

$$\mathbf{N} = \begin{pmatrix} 3 & 2 & 10 \\ 2 & 1 & 0 \\ 4 & 3 & 15 \end{pmatrix}.$$

(Matrices)

Matrix addition

f we want to see how much of each type of bread was supplied on each of the three days during weeks five and six, we can add together two matrices.

$$\mathbf{A} + \mathbf{N} = \begin{pmatrix} 10 & 4 & 30 \\ 8 & 3 & 0 \\ 12 & 5 & 40 \end{pmatrix} + \begin{pmatrix} 3 & 2 & 10 \\ 2 & 1 & 0 \\ 4 & 3 & 15 \end{pmatrix} = \begin{pmatrix} 13 & 6 & 40 \\ 10 & 4 & 0 \\ 16 & 8 & 55 \end{pmatrix}.$$

The logic of the problem dictates how the *elements* (that is the numbers that make up the matrices) combine. This turns out to be a useful definition to adopt for *matrix addition*. The rule leads easily to the rule for multiplying a matrix by an ordinary number. For instance,

$$5\mathbf{A} = 5 \times \begin{pmatrix} 10 & 4 & 30 \\ 8 & 3 & 0 \\ 12 & 5 & 40 \end{pmatrix} = \begin{pmatrix} 50 & 20 & 150 \\ 40 & 15 & 0 \\ 60 & 25 & 200 \end{pmatrix}.$$

Compare the baker's order for the first five weeks.

Matrix multiplication

The idea of how to multiply matrices arises from extending the problem. A large white loaf costs 15p, a small brown costs 9p and a roll 2p, let us suppose. This information can be given in a table, like this:

Type of loaf	Price in pence
Large white	15
Small brown	9
Roll	2

Notice that we keep the types of loaf, and hence the prices, in the same order down the column as in the previous matrices across the row.

This provides a new matrix, **P**.

$$\mathbf{P} = \begin{pmatrix} 15 \\ 9 \\ 2 \end{pmatrix}.$$

This matrix has three rows and one column (3 X 1) whereas **A** and **N** have three rows and three columns (3 X 3). We set this price matrix out in a column because we have a convention that, if we are going to combine two matrices in the way that we shall call *matrix multiplication* we always combine a row of the left-hand matrix with a column of the right-hand one.

If we want to find the cost of Monday's bread in the sixth week, we have to put together information from the matrices **N** and **P**. The order for white, brown and rolls was (3 2 10) and the prices, in correct order $\begin{pmatrix} 15 \\ 9 \\ 2 \end{pmatrix}$

Working in pence, 3 white at 15 cost 45, 2 brown at 9 cost 18, 10 rolls at 2 cost 20 total cost 45 + 18 + 20 = 83. Written in matrix form:

$$(3 \quad 2 \quad 10) \begin{pmatrix} 15 \\ 9 \\ 2 \end{pmatrix} = (45 + 18 + 20) = (83).$$

We want the result of matrix X matrix to be a matrix, that is why the answer, 83, is written in a bracket. It is itself a 1 X 1 matrix. It is interpreted to mean a total cost of 83p.

For Wednesday of week six the order was different, 2 white, 1 brown, no rolls, or (2 1 0). To get the total we work:

$$(2 \quad 1 \quad 0) \begin{pmatrix} 15 \\ 9 \\ 2 \end{pmatrix} = (30 + 9 + 0) = (39)$$

The calculation for Friday is $(4 \quad 3 \quad 15) \begin{pmatrix} 15 \\ 9 \\ 2 \end{pmatrix} = (60 + 27 + 30) = (117).$

As the matrix $\begin{pmatrix} 15 \\ 9 \\ 2 \end{pmatrix}$ occurs each time, we can condense our work and write:

$$\begin{pmatrix} 3 & 2 & 10 \\ 2 & 1 & 0 \\ 4 & 3 & 15 \end{pmatrix} \begin{pmatrix} 15 \\ 9 \\ 2 \end{pmatrix} = \begin{pmatrix} 83 \\ 39 \\ 117 \end{pmatrix}.$$

Note how the rows 'dive' into the columns, one by one, to provide the three totals in the resulting matrix.

In the seventh week the prices were increased by 2p for both the white and the brown loaf. The new price matrix is therefore

$$\begin{pmatrix} 17 \\ 11 \\ 2 \end{pmatrix}.$$

If the baker's order remains the same as in the previous week, we find the costs for Monday, Wednesday and Friday by multiplying again in a similar manner:

$$\begin{pmatrix} 3 & 2 & 10 \\ 2 & 1 & 0 \\ 4 & 3 & 15 \end{pmatrix} \begin{pmatrix} 17 \\ 11 \\ 2 \end{pmatrix} = \begin{pmatrix} 93 \\ 45 \\ 131 \end{pmatrix}.$$

Check that 93 comes back from 3 × 17 + 2 × 11 + 10 × 2. How were 45 and 131 obtained? (See Answer 46 at the back of the book.)

Once again we can condense the work by writing the multiplying matrix once for the calculations for weeks six and seven:

$$\begin{pmatrix} 3 & 2 & 10 \\ 2 & 1 & 0 \\ 4 & 3 & 15 \end{pmatrix} \begin{pmatrix} 15 & 17 \\ 9 & 11 \\ 2 & 2 \end{pmatrix} = \begin{pmatrix} 83 & 93 \\ 39 & 45 \\ 117 & 131 \end{pmatrix}.$$

This is a bigger and better example of matrix multiplication, a 3 × 3 matrix multiplying a 3 × 2 matrix. The same thing can be extended to two matrices of any size, provided a dimension of each is the same. Which are these dimensions? (See Answer 47.) If they are not the same, then the matrices cannot be multiplied. As is common in algebra, we do not use a multiplication sign between matrices.

Please note that, in this particular example, it would make further sense to add together, say 83 + 39 + 117. What would this give? However this is *not* a general property of matrices, it is just because of the meaning of the numbers in this case.

Try the following examples for yourself. Beware, two of the examples are impossible. When you have tried them, check with Answer 48.

(a) $\begin{pmatrix} 1 & 2 \\ 2 & 3 \end{pmatrix} \begin{pmatrix} 2 \\ 4 \end{pmatrix}$

(b) $\begin{pmatrix} 1 & 2 \\ 2 & 3 \end{pmatrix} (2 \quad 4)$

(c) $\begin{pmatrix} 1 & 2 \\ 4 & 3 \end{pmatrix} + \begin{pmatrix} 2 & 3 \\ 5 & 1 \end{pmatrix} + \begin{pmatrix} 4 & 5 \\ 3 & 0 \end{pmatrix}$

(d) $\begin{pmatrix} 2 & 3 & 1 \\ 1 & 0 & 10 \end{pmatrix} + 5 \begin{pmatrix} 5 & 4 & 2 \\ 0 & 1 & 1 \end{pmatrix}$

(e) $(4 \quad 1) \begin{pmatrix} 2 & 2 \\ 0 & 5 \end{pmatrix}$

(f) $(2 \quad 6 \quad 4) + \begin{pmatrix} 1 \\ 3 \\ 5 \end{pmatrix}$

(g) $\begin{pmatrix} 1 & 0 \\ 0 & 1 \end{pmatrix} \begin{pmatrix} 4 & 5 \\ 6 & 7 \end{pmatrix}$

Matrix algebra has a number of interesting features. One of them is this. In the arithmetic of numbers 3 + 4 = 4 + 3 = 7. You may remember that we describe this by saying addition is *commutative* when we are dealing with numbers. Will it be true that the addition of matrices is commutative?

Consider two simple matrices, for example

$$A = \begin{pmatrix} 2 & 4 \\ 3 & 1 \end{pmatrix} \text{ and } B = \begin{pmatrix} 1 & 3 \\ 2 & 4 \end{pmatrix}.$$

Work out A + B and B + A; what do you find? (See Answer 49.)

Work out also $A \times B$ and $B \times A$. This is what happens:

$$A \times B = \begin{pmatrix} 2 & 4 \\ 3 & 1 \end{pmatrix} \begin{pmatrix} 1 & 3 \\ 2 & 4 \end{pmatrix} = \begin{pmatrix} 10 & 22 \\ 5 & 13 \end{pmatrix}$$

$$B \times A = \begin{pmatrix} 1 & 3 \\ 2 & 4 \end{pmatrix} \begin{pmatrix} 2 & 4 \\ 3 & 1 \end{pmatrix} = \begin{pmatrix} 11 & 7 \\ 16 & 12 \end{pmatrix}$$

In this case it is certainly not true that $A \times B$, or AB is equal to BA; nor is it generally. The algebra of matrices has an operation of multiplication which is non-commutative and can be contrasted with the algebra of numbers in which multiplication is commutative.

An application to geometry

There is a close link between a matrix and a vector; indeed the word vector is often used for a matrix with two rows and only one column like this: $\begin{pmatrix} 5 \\ 2 \end{pmatrix}$. You will remember that this is the way in which we describe a translation of 5 units in the x-direction and 2 in the y-direction. Fig. 12.1 shows the vector $\begin{pmatrix} 5 \\ 2 \end{pmatrix}$. If we multiply this by the matrix $\begin{pmatrix} 0 & 1 \\ 1 & 0 \end{pmatrix}$ we get:

$$\begin{pmatrix} 0 & 1 \\ 1 & 0 \end{pmatrix} \begin{pmatrix} 5 \\ 2 \end{pmatrix} = \begin{pmatrix} 2 \\ 5 \end{pmatrix}.$$

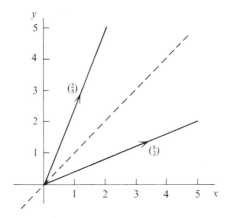

Fig. 12.1

The vector $\begin{pmatrix} 2 \\ 5 \end{pmatrix}$ is also shown in the diagram. Can you see the geometrical relation between these vectors? The broken line is intended to be helpful.

The matrix $\begin{pmatrix} 0 & 1 \\ 1 & 0 \end{pmatrix}$ reflects all vectors in the broken line, whose equation is $y = x$.

If you worked through the last set of examples, as we hope that you did, you should have remarked the very special result of multiplying by $\begin{pmatrix} 1 & 0 \\ 0 & 1 \end{pmatrix}$. You will also see why it is called the *identity matrix* for multiplication.

With purchase and prices matrices it does not make much sense to allow the numbers that are their elements to be negative. With geometrical matrices and others this makes good sense. Multiply the vector $\begin{pmatrix} 5 \\ 2 \end{pmatrix}$ by the two matrices,

$$(a) \ \begin{pmatrix} 1 & 0 \\ 0 & -1 \end{pmatrix}; \quad (b) \ \begin{pmatrix} -1 & 0 \\ 0 & -1 \end{pmatrix}.$$

Plot the vectors into which they transform $\begin{pmatrix} 5 \\ 2 \end{pmatrix}$ and so find what transformations these matrices represent. (See Answer 50.)

An application to algebra

You may remember that two equations in two unknowns, x and y, which are true at the same time, are called *simultaneous equations* and a typical pair look like this:

$$\left. \begin{array}{l} 2x + y = 4 \\ 3x + 2y = 5 \end{array} \right\}.$$

Using the notion of matrix multiplication we can write the two equations with just one equality sign, like this:

$$\begin{pmatrix} 2 & 1 \\ 3 & 2 \end{pmatrix} \begin{pmatrix} x \\ y \end{pmatrix} = \begin{pmatrix} 4 \\ 5 \end{pmatrix}.$$

Now we are going to do what no teacher of modern mathematics should do! That is, to produce a matrix solution, like a rabbit out of a hat, without preliminary exercise, motivation, or even explanation. Remember that 'solution' means finding the values of x and y for which the two equations are true. Remember also that the two sides of an equation remain equal if one does the same (allowable) thing to both sides. Such an allowable thing is to multiply both sides by another matrix, like this:

$$\begin{pmatrix} 2 & 1 \\ 3 & 2 \end{pmatrix} \begin{pmatrix} x \\ y \end{pmatrix} = \begin{pmatrix} 4 \\ 5 \end{pmatrix}$$

and so $\begin{pmatrix} 2 & -1 \\ -3 & 2 \end{pmatrix} \begin{pmatrix} 2 & 1 \\ 3 & 2 \end{pmatrix} \begin{pmatrix} x \\ y \end{pmatrix} = \begin{pmatrix} 2 & -1 \\ -3 & 2 \end{pmatrix} \begin{pmatrix} 4 \\ 5 \end{pmatrix}.$

Now we multiply the left-hand pair together. Check that this gives

$$\begin{pmatrix} 1 & 0 \\ 0 & 1 \end{pmatrix} \begin{pmatrix} x \\ y \end{pmatrix} = \begin{pmatrix} 2 & -1 \\ -3 & 2 \end{pmatrix} \begin{pmatrix} 4 \\ 5 \end{pmatrix}.$$

Now multiply both sides out completely:

$$\begin{pmatrix} x \\ y \end{pmatrix} = \begin{pmatrix} 3 \\ -2 \end{pmatrix}.$$

The only way in which two matrices can be 'equal' is for them to have the same elements in the same places throughout. This means that $x = 3$ and $y = -2$. The equations are solved! Check by putting 3 for x and -2 for y in the original equatic

We have shown you this slick way of solving simultaneous equations for two reasons. The first is that your children may be using this method, and you now know *what* they are doing, if not why they are doing it; and it is an elegant methoc The second is that this method, and modifications of it, can be used to solve numbers of similar equations, in dozens of unknowns, $x, y, z, t, \ldots$, when they arise, a they do, in practical problems in engineering, economics, optimisation and so on. also provides a preview of important applications yet to come.

An application to networks

We mentioned networks in Chapter 8, and earlier in this chapter said that networks provided an alternative way into the idea of a matrix. Let us take the topic a little further and study a matrix that is different from the arithmetic matrices (purchase and price), the transformation matrices and the matrices obtained from equations that we have glimpsed so far.

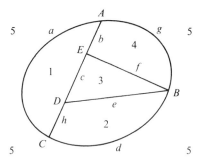

Fig. 12.2

Figure 12.2 shows a network consisting of five nodes, A, B, C, D, E, joined by eight arcs, a, b, c, d, e, f, g, h and dividing the plane of the paper into five regions, 1, 2, 3, 4, and the outside, 5.

We did enough work in the geometry section to know that it is not always easy to tell whether networks are equivalent or not. When they get more complicated it becomes important to find some systematic way of describing them. We do this by compiling matrices which show how nodes lie on arcs, how arcs divide regions, etc. As with loaves of bread and days of the week, we start with a table, and later omit the headings.

We shall start with the table for 'nodes onto regions'. This means that nodes go down the side, labelling the rows, and regions across the top, labelling the columns:

Regions

	1	2	3	4	5
A	1	0	0	1	1
B					
Nodes *C*					
D					
E					

The node *A* is the point where regions 1, 4 and 5 meet. In the row labelled *A* we have put 1s under 1, 4 and 5 and 0s under 2 and 3, these being the regions which do not meet at *A*. Similarly the row for *C* is (1   1   0   0   1). Fill in the remaining rows for yourself.

Omitting the headings and inserting brackets we have what is called an *incidence matrix*. You should get this:

$$\mathbf{M} = \begin{pmatrix} 1 & 0 & 0 & 1 & 1 \\ 0 & 1 & 1 & 1 & 1 \\ 1 & 1 & 0 & 0 & 1 \\ 1 & 1 & 1 & 0 & 0 \\ 1 & 0 & 1 & 1 & 0 \end{pmatrix}.$$

Try to complete the matrix **N** which describes how regions relate to arcs. Start with a table with regions down the side (labelling the rows), and arcs along the top (labelling columns). Try to decide, before you start, the dimensions that you will expect for the resulting matrix. Form the matrix product **MN**. The result is connected with the incidence matrix **P** for nodes onto arcs. Can you see how? We should warn you that this is probably the hardest exercise in the book! As always, the answer is in the back (Answer 51).

Networks are not purely abstract. For instance, they can be used to study traffic circulation and the timing of trains, and for the steps of a complicated manufacturing operation, with the time taken for each. Transforming the resulting matrices can help to solve problems, real problems in the everyday world, particularly now that the existence of large computers means that huge arrays of numerical data can be scanned and manipulated very quickly.

This is far from the end of matrices and their application, even so far as the secondary course is concerned. It is enough, we hope, to show you what they can do and the versatility of the method. There are a number of books on the market, of all degrees of difficulty. You will also find constant references and examples throughout SMP texts, which will help you if you want to go any deeper.

13 Evaluation: is new mathematics here to stay?

Evaluation

The object of the recent changes in syllabus content, as we have said before, is to produce men and women with a better understanding of mathematics at all levels, and with more ability to use it. The test of whether the movement has been success ful or not is therefore not merely a matter of looking at examination results. In fac the pass rate for public examinations in the United Kingdom, A-level, O-level and CSE, is governed by the percentage of entrants that policy dictates shall be permitted to pass. In the early days of SMP, for instance, it was a decision of the examination board concerned that the percentage passing on the new papers was to be pegged at the percentage passing the old. Examination pass-rate statistics are therefore irrelevant, nor is it possible to say that some particular grade implies an understanding of mathematics, whereas another does not.

We hope it does not sound as though we are evading the issue! Having written the word 'evaluation' into our final chapter title you may be disappointed that our findings are largely subjective.

Subjective evaluation is far from useless. The trained teacher knows whether a subject is liked and whether ideas are being absorbed, discoveries made. Throughou the years in which the new work has been spreading, children have been quizzed and observed continuously. In one investigation, carried out in 1962, we found tha boys of eleven rated mathematics among the most popular of subjects on entry to their secondary schools. Only a small proportion of boys in the same schools thought the same at the age of sixteen. This was before new mathematics had been introduced. Subsequent investigations have suggested that views have changed. It would be interesting to know what teachers in secondary schools of differing types find to be the present attitude.

An essential criterion for putting new material into the curriculum is whether it is interesting. You will now be in a good position to say whether you find the work that we have done interesting or not! Children, of course, form a captive audience from the beginning of schooling until such time, usually after the age of sixteen, when mathematics becomes a voluntary subject. Then they can express their views by their choice of subject.

In his inaugural address as Professor of Mathematics in the University of Southampton in May 1961, Dr Bryan Thwaites warned of the implications for the future teaching of mathematics at all levels of the shortage of mathematicians enter

[156]

ing the teaching profession. It was largely due to the inspiration of that address that SMP got under way. Another of the stated aims was, through making the subject-matter intrinsically more interesting, to induce more ordinary mortals to take up the subject at A-level and subsequently at higher level. In our young days mathematics was thought of as being mainly for the genius! This attitude had to be changed.

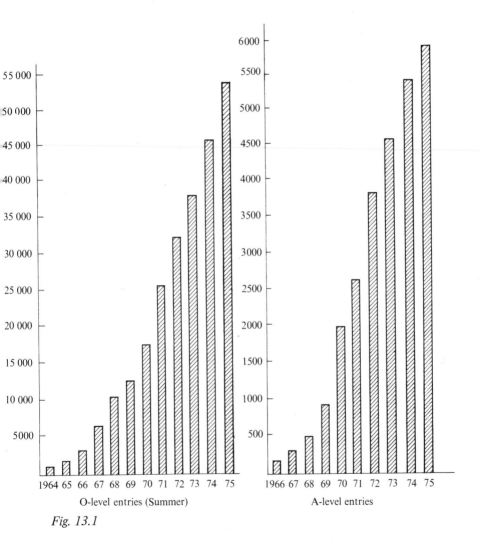

Fig. 13.1

The attitude has been changed. Each year the numbers of children taking mathematics at A-level has increased, see Fig. 13.1, and within this overall increase, the numbers of those taking the A-level papers set on the SMP syllabus has risen from

1182 in 1969 to 5371 in 1974. (In the same period the O-level entries on the SMP syllabus rose from 12 879 to 45 052.) The graphs illustrate the increasing contribution made by SMP to the teaching of mathematics in England.

Each year, too, the number of graduates in mathematics has increased, so that now the number entering universities and other institutions of higher education to read mathematics is second only to medicine in total. To this extent, and inasmuch as this increase has accompanied the changes in syllabus, we can find grounds for satisfaction. However, a general increase in numbers staying on for sixth-form education, together with the recent raising of the school leaving age, previously 15, to 16, means there are still not enough trained teachers of the subject in secondary schools. Dr Thwaites's warning still holds true; there is grave danger that the teaching of the subject may enter a descending spiral.

We are on more difficult ground still when we come to testing the understanding of fundamental mathematical ideas, and try to judge whether our objective of better understanding is being achieved. The main problem is to obtain agreement on what constitutes mathematical understanding. It is difficult even to define mathematical skills, as a recently published booklet on the subject makes clear.† Modern British projects have been criticised for not themselves engaging in systematic testing of attitudes, skills and understanding, but they have neither the resources nor the expertise for doing this. Fundamental research is being carried out, however, in the United Kingdom by the Schools Council and by the National Foundation for Educational Research, and by a number of bodies in the United States, such as Educational Testing Services, Inc., of Princeton.

Teacher attitudes to new mathematics

Out of 6500 secondary schools in England and Wales, it has been estimated that 4000 were engaged in some sort of new mathematics in 1972. In Scotland, virtually all secondary schools have adopted their own version of the new syllabuses. It would be quite impossible for this to have happened if teachers had not been in favour. Of course the change-over has followed a predictable pattern. A small group of pioneers had the ideas. Keen teachers, not always the youngest ones, quickly attended conferences and courses and returned to their own school determined to take the plunge. As the movement spread, it became 'the done thing', and pressure for change built up among head teachers, mathematics advisers to local education authorities, governors and parents. Sometimes changes were made prematurely, sometimes they were more in the letter than the spirit. There now remains a sizeable minority-group of schools and teachers who have clung to traditional syllabuses, sometimes through conviction, sometimes through ignorance of what the new has to offer, sometimes through inertia. We entitled this chapter, 'Is new mathematics

---

† *Manipulative Skills in School Mathematics*, published by the School Mathematics Project from Westfield College, London NW3 7ST, obtainable on request.

ere to stay?' In the sense that there will be no return to the old, there can be little doubt that the answer to this is yes.

The new ideas continue to spread through conferences, journals and through the work of the Teachers' Centres. The major projects organise conferences on a national scale. In 1974, SMP alone held conferences in Nottingham, Exeter, Lancaster, Huddersfield, Cambridge. Ponteland, Godalming and London. Many thousands of teachers have attended over the years. The principal mathematical journals for secondary teachers are the *Mathematical Gazette, Mathematics in School* and *Mathematics Teaching* in the United Kingdom, and the *Mathematics Teacher* in the United States. Between them they form an invaluable guide to what is going on in the schools.

Most Education Authorities have now set up Teachers' Centres in their areas. These are houses, schools or purpose-built centres with facilities for working groups, study and lectures, mostly with a library and many with resident or full-time wardens. Mathematics is a popular subject of study as might be expected at a time of change.

The enthusiasm of teachers for new syllabuses is not confined to those in the United Kingdom and the United States. The movement is truly international, although it has taken rather different forms in different countries. A quick glance at a recent SMP Director's Report reveals that teachers from the United Kingdom visited Zambia, Malawi, Kenya, the Canary Islands, Swaziland, Botswana, South Africa, Malta, Sri Lanka, Tanzania, Ghana, Nigeria and the West Indies in the space of only a year or two. British texts are being adapted for use abroad in Australia, South Africa, Holland, Germany, the United States, Norway and Sweden, and probably in others. Foreign language editions are being prepared in Arabic, French, Italian and Turkish. The keen mathematics teacher with a desire for travel finds plenty of opportunities.

Examinations

What then of examinations? Passes in mathematics are still needed. Children must not only achieve understanding, but obtain the pieces of paper which purport to give evidence of this (however unsure that evidence may be). New examinations have been provided for the main projects and have been sponsored by the Examination Boards. The earliest O-levels were set in 1964, A-levels in 1966 and CSE soon after. They are intended to be comparable in difficulty and prestige to the traditional ones. It is not possible to tell from most certificates which syllabus a candidate has been tested on, since it will simply say 'mathematics'.

The examinations themselves have changed with the syllabus. Some are 'Mode 3', that is to say they consist of two parts, one common to many schools and one peculiar to the particular school and run by the school's own staff to the school's own syllabus and requirements. In some subjects, project material and investigations carried out by the individual child can be submitted to a Board, in addition to conventional examinations. This is not yet common practice with mathematics, but is a natural consequence of the newer types of teaching described in Chapter 2, and we may see it coming in more and more.

Even the examination questions are of a new kind. A typical O-level question of a traditional kind is this. 'A cylindrical jar of radius 1.75 centimetres contains water to a depth of 4 centimetres. Find the increase in depth of water when a steel ball of radius 0.75 centimetres is completely immersed in it.' The most testing part is perhaps appreciating what steps are needed. An average-to-good candidate may take a few minutes to see that he needs to divide the volume of the steel ball by the cross-sectional area of the jar to obtain the increase in depth. This is where his understanding of volume, of its relation to area, the concept of conservation, ideas of ratio, etc. will be tested. The computation of the answer now involves substituting the formula for the volume of a sphere, and for the area of a circle, and then dividing the results. Using (traditionally) logarithms this may take three or four times as long as the necessary mathematical decision and understanding, that is, the bulk of the time will have been spent on routine processes of computation. While it is obviously necessary to test that a candidate can perform calculations accurately, the main purpose of the examination will be to test mathematical ability, and thus the balance of time spent is the wrong way round.

Two things can be done. The child can be equipped with a pocket calculator; but this will be expensive and may not occur widely in the near future. The other is to alter the style of question so that more mathematical decisions are required at the expense of computation. One way of doing this is to change to the so-called *multiple choice* type of question. Keeping to the same topic, a cylindrical jar and a steel ball, here are some typical questions.

Tick the correct alternative answer to the following questions.

1. For the ball to be completely submerged the depth of water must be:
   (*a*)   not more than 1.75 centimetres;
   (*b*)   exactly 1.5 centimetres;
   (*c*)   at least 1.75 centimetres;
   (*d*)   none of these answers is necessarily correct.

2. The change in the depth of water is given by the fraction:
   (*a*) $\dfrac{\text{area of ball}}{\text{area of cross-section}}$;
   (*b*) $\dfrac{\text{radius of ball}}{\text{radius of jar}}$;
   (*c*) $\dfrac{\text{area of cross-section}}{\text{volume of ball}}$;
   (*d*)   none of these answers is correct.

3. The effect on the change of depth of substituting a ball of twice the radius would be to:

'The effect . . . of . . . a ball of twice the radius'

- (*a*)   double it;
- (*b*)   multiply it by 1.75^2;
- (*c*)   quadruple it;
- (*d*)   none of these is correct.

4. The formula for the volume of a spherical ball of radius r is $\frac{4}{3}\pi r^3$. The volume of the ball of radius 0.75 centimetres will be:

- (*a*)   less than 1.5 cubic centimetres;
- (*b*)   less than 4 cubic centimetres;
- (*c*)   more than 4 cubic centimetres;
- (*d*)   there is not enough information to be able to calculate this.

Decisions about variation, geometry, ratio, etc., confront the child in these questions. Future research will tell whether the newer types of paper are better or worse than the old ones. Some teachers say that it is important to test the ability of a child to think through a long piece of calculation as well as to answer a number of short questions and some boards set papers of both types. The multiple-choice type of question may detect mathematical ability as well as, if not better than, the conventional type of paper. It has the added advantage of being very much easier to mark.

Further education and university entrance

One stern test of new mathematics is already taking place. Children brought up on the new projects are applying for entrance to the universities and polytechnics. Mathematics departments have largely welcomed the new work, but there have been sounds of disapproval from some scientists and engineers. Regrettably, some of those who have been most critical of the effect of the new examinations and/or the new syllabuses have not taken very much trouble to see what, in fact, they contain. As university teachers receive more of the new generation into their classes, we can look forward to more outspoken and informed comment on the value and efficiency of the new syllabuses and the new methods of teaching.

It must not be forgotten that there have been changes too in the content of university courses in mathematics, science and technology. Reliability engineering and quality control call for probability theory. Transformation methods are being replaced by methods derived from abstract algebra. No department is now without access to computers, and the sort of analysis which is common depends on ideas of sets and structure, even if no previous acquaintance with the ideas of programming is demanded. Matrix and vector methods also are in general use. For us there is no doubt that children leaving the schools are better prepared to go on to many branches of education involving mathematics than they were formerly.

Changes in the teaching of science

It is no accident that the revolution in mathematics teaching coincided with the Nuffield Science Teaching Project, which transformed much of the science teaching. Their aims are in broad agreement with those of the mathematical projects, and the root causes of dissatisfaction with the older ways of teaching are also very similar. In a survey carried out by the Schools Council as long ago as March 1968, it was estimated that at least 1000 schools in England and Wales were using Nuffield material fully or partially, and that the cost of this has been in excess of £1 000 000 the number now must be far greater. Perhaps more indicative of the spread is the fact that in a recent survey of 120 Local Education Authorities, only 12 had any schools in their area which were not concerned with Nuffield Science in some way or another. Mathematicians and scientists are aware that, in their eagerness to get to grips with their own problem, they have not always consulted each other as fully as they should have done, but steps are now being taken to achieve more co-ordination

The reaction of employers

Employers have sometimes been too ready to grumble about the mathematical equipment of school-leavers. The human animal is an imperfect calculating machine and will always make mistakes. As offices and workshops become more automated, more understanding of how machines work will be essential. However, back-of-the envelope calculations are still required. At the present time there is an understandable tendency to blame new methods when the calculations do not always turn out to be correct! However, no syllabus change can eliminate silly mistakes and minor inaccuracies. Some employers agree that the new ways are producing people who are more flexible, more willing to think out a problem from first principles. One of our difficulties, in the schools, has been to keep employers informed about what we are trying to do. This was one of the reasons why we wrote this book.

What, after all, is mathematics?

In our introductory chapter we said that we wanted to give each child some idea of the answer to this, very difficult, question. We hope you feel that you have a better appreciation of the totality and unity of mathematics as a result of reading this short review.

Again, but for the final time, we turn to examples to help explain what we mean. Consider the statement 'All men are less than 10 feet tall.' It is almost certainly true, but is not absolutely true. Its truth derives from experience, experiment. Its force could be increased by arguments about the necessary breadth of leg to support the proportionately enormous body-weight, etc. In logical form it says that the truth of statement A, 'He is a man' implies the truth of statement B, 'He is less than 10 feet tall.' This implication is not, in logic, 'true'.

Compare the statement 'There is only one even prime number.' A prime number is defined to be a number whose only factors are 1 and itself. Any even number, other than 2, has 2 as a factor in addition to 1 and itself; it is therefore not prime. The truth of statement A, 'The number is an even prime' implies absolutely the truth of statement B, 'The number is 2.' This implication is incontestably 'true'.

The difference between these two propositions lies at the heart of the difference between mathematics and science, or indeed any other discipline. Mathematics is about certainties. A number, a factor of a number, division, the existence of 2 are all axioms or operations defined unambiguously and independently of any system of experiment. It is not logically important that numbers occur in the real world, that division, of a sort, can be done with a cake knife. Mathematicians know what they are talking about. What they base their statements on will be the same in the future as in the past. Not for them will there be new theories supplanting old theories to come nearer to observable 'truth'. Mathematical truth is absolute.

An enigmatic definition of mathematics is 'The class of propositions A implies B.' This is not the whole truth, but it is a good starting point. Whenever you assert absolute truth you are doing mathematics of a sort. This definition has a further point to make. Mathematics is about relations, and implication is a type of relation. The common thread which runs through arithmetic, algebra, geometry, statistics, trigonometry, etc. is that the elements that they deal with are in some relation to one another. These relations ($=, >$, etc.) are studied as well as the elements to which they refer, as we have seen.

The certainty implicit in mathematics is both its glory and, for some, a drawback. Children and students alike enjoy this certainty, the intricacy, depth and self-sufficiency of mathematics. But they often regret the absence of controversy, of humanity. Humanity can enter through applications, hence there is in the United Kingdom a very strong tradition of applying the subject. This is true not only in the schools of mathematics at universities and polytechnics, but in all schools, even at infants' level.

This is where we started, with the links between the steely elegance of mathematics and the functional brilliance of what mathematics can do for humanity. Though mathematics does not change, it develops all the time. It is a bold man who will prophesy what the mathematical requirements for our children will be in ten years' time, much less fifty. One thing is sure, they will be different. Thus the best mathematical training is the most adaptable. So our final answer to whether the new mathematics is here to stay is NO! In ten years' time it will be the old mathematics and will have to be looked at again, reshaped, re-emphasised and taught in still newer ways.

The end − or the beginning?

If your curiosity has been aroused by any of the topics discussed in this book you may find interest and pleasure in the SMP series of books published by Cambridge University Press. These are eminently readable, always remembering that 'readable' in the context of a book on mathematics means having pencil and paper to hand and working through the introductory material before plunging into the exercises.

Up to O-level there are two series, one for those who like to speed along, Books 1−5, and another series for those who like a more leisurely pace, Books A−H and X, Y, Z. The mathematics in Books A−D is also published as packs of cards especially designed for use with small groups of pupils of mixed ability. Should your enthusiasm take you beyond O-level, there are 4 books to A-level.

Each of the O-level books has its Teacher's Guide which contains much background material as well as answers. The A-level books also have books of answers and hints.

Beyond A-level there are five books which form the basis of the work for the Further Mathematics examination and each of these concentrates on one or two related topics.

Answers and hints to problems in the text

Chapter 1

Answer 1

In Fig. 1.1, the sum of every row, column and diagonal is 15.

Answer 2

In Fig. 1.2, notice that except for the 1 at the beginning and end of each line, each number can be obtained from the previous line by adding the number above it to the number on its left, for example, the 6 in the last horizontal row is 5 + 1; 15 = 10 + 5, 20 = 10 + 10 etc. so the next horizontal row is 1, 7, 21, 35, 35, 21, 7, 1. The total for this row is 128. The column totals are now 8, 28, 56, 70, 56, 28, 8, 1. These are not all divisible by 8! But they do provide the entries in the next line, displaced one to the left.

    The Fibonacci numbers, 1, 1, 2, 3, 5, 8, 13, . . . , are so called because Leonardo Fibonacci (1175–1230) carried out a great deal of research into their properties. In order to continue the sequence note that each number is the sum of the two previous numbers.

    See also Chapter 6 of *SMP Book 2*.

Chapter 4

Answer 3

The next prime is 11. To obtain 11 as the sum of two smaller primes is impossible, 7 + 5 is too big and 7 + 3 too small.

    The smaller primes may be expressed as the difference of two other primes, for example, 2 = 5 − 3, 3 = 5 − 2, but for larger primes only those that differ by two can be expressed as the difference of two primes, 17 = 19 − 2, 29 = 31 − 2, 41 = 43 − 2, etc. All primes except 2 are odd, so adding or subtracting pairs of primes will always give an even number unless 2 is one of the pair being added or subtracted.

Answer 4

A theory can be considered to be true until it is proved untrue in a single instance. No particular number of successes amounts to 'certainty', however, for not every case can be tried, unless there are only a fixed number of possibilities.

Answer 5

$n^2 + n + 41$ and $n^2 − n + 41$ will fail to give primes when n has any value which is a multiple of 41, for example, 82, 123, etc.

Answer 6

Using the formula $4^n + 3$,
$n = 1$ gives $4 + 3 = 7$ (prime),
$n = 2$ gives $4^2 + 3 = 16 + 3 = 19$ (prime),
$n = 3$ gives $4^3 + 3 = 64 + 3 = 67$ (prime),
$n = 4$ gives $4^4 + 3 = 256 + 3 = 259$,
             but $259 = 7 \times 37$, so is not prime.
See also Chapter 6 of *SMP Book 1*.

Chapter 5

Answer 7

$O + E = E + O = O$

Answer 8

(a) E,   (b) O,   (c) O.
You may have noted that it is only necessary to count up the Os.

Answer 9

(a) $997 + 1998 + 3 = (997 + 3) + 1998$ (addition of numbers is associative)
$= 1000 + 1998$
$= 2998$.
(b) $\frac{1}{4} \times \frac{1}{3} \times 20 \times 30 = (\frac{1}{4} \times 20) \times (\frac{1}{3} \times 30)$ (using the associativity of numbers under multiplication)
$= 5 \times 10$
$= 50$.
For commutativity and associativity, see Chapter 4 of *SMP Book 2*.

Answer 10

+	0	2
0	0	2
2	2	4

The algebra of 0 and 2
is not closed under addition

For closure and operation tables see Section 2, Chapter 11 of *SMP Book 3*.

Answer 11

(a) $A \cap B = \{O, N\}$, $A \cup B = \{A, D, H, L, M, N, O, P, S, T, U\}$;
(b) $R = \{2, 4, 6, 8, 10, 12, 14, 16, 18, 20\}$,
$R \cap F = \{6, 12, 18\}$,
$R \cup F = \{2, 3, 4, 6, 8, 9, 10, 12, 14, 15, 16, 18, 20\}$;
(c) $M' = \{$all types of metal except gold, silver, lead, copper$\}$,
$M' \cap L = \{\ \}$ or $\varnothing$, the empty set;
(d) $D' = \{$even numbers less than 20$\}$,
$D' \cap E = \{\ \}$ or ○,
$E' = \{1, 2, 4, 6, 8, 10, 11, 12, 13, 14, 15, 16, 17, 18, 19\}$,
$D \cup E' = \{1, 2, 3, 4, 5, 6, 7, 8, 9, 10, 11, 12, 13, 14, 15, 16, 17, 18, 19\}$;

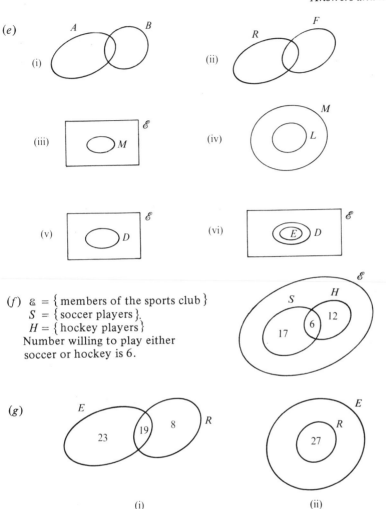

(e)

(i)

(ii)

(iii)

(iv)

(v)

(vi)

(f)   ℰ = { members of the sports club }
S = { soccer players }.
H = { hockey players }
Number willing to play either
soccer or hockey is 6.

(g)

(i)

(ii)

The smallest number of houses
getting both bills on the same
day is 19.

The largest number of houses
getting both bills on the same
day is 27.

In SMP work, sets are introduced in *Book 1*, Chapter 2 and continued in Chapter
11, *Book 3*. The usefulness of sets is so great that reference is often made to them
in other chapters too.

Chapter 6

Answer 12

(a) $7\frac{1}{2}$ min;   (b) $12\frac{2}{5}$ min;   (c) $2\frac{4}{5}$ min.

(c) is likely to be the least accurate because the time is influenced to some extent
by the temperature of the stove itself, for example, if saucepans are being heated on
the top it may be quite warm before the oven is put on. Another reason is the big
gap between the first two readings resulting in a line whose position can only be
guessed.

(*d*) just over 130°C;
(*e*) approximately 375 °C.
See linear relations, Chapter 11, *SMP Book 1*.

Answer 13. Fig. 6.3

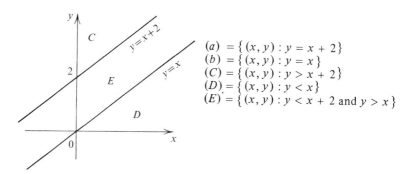

$$(a) = \{(x, y) : y = x + 2\}$$
$$(b) = \{(x, y) : y = x\}$$
$$(C) = \{(x, y) : y > x + 2\}$$
$$(D) = \{(x, y) : y < x\}$$
$$(E) = \{(x, y) : y < x + 2 \text{ and } y > x\}$$

Answer 14

$2y = 5x - 5$ becomes $2 \times 5 = 5 \times 3 - 5$, which is true.

Answer 15

$5 = 3 + 2$; 5 is not *greater* than $3 + 2$.

Answer 16

Fig. 6.5: (*a*) is graph of $y = 2x + \frac{1}{2}$;
          (*b*) is graph of $y = 2 - x$.

Answer 17

Graphs can only give approximations, a solution that looks like $x = 1$ might be $x = 0.99$.

Answer 18

Identities (*a*), (*c*), (*d*), (*e*), (*f*). Equations (*b*), (*g*).

Answer 19

Under multiplication, 1 is the identity element in the set of numbers.

Answer 20

The equation $2x + 3 = 8$.
    Under addition of numbers the inverse element of $+3$ is -3.
    Add -3 to both sides:

$$2x + 3 + -3 = 8 + -3,$$
$$2x + 0 \qquad = 5.†$$

0 is the identity element under addition so does not change the $2x$,

$$2x \qquad = 5.$$

The inverse of 2 under multiplication is $\frac{1}{2}$.

---

† If you are a purist reader we should perhaps add that at this point we have also made use of the associativity of addition in asserting that $(2x + 3) + -3 = 2x + (3 + -3)$.

Multiply both sides by $\frac{1}{2}$:

$$\tfrac{1}{2} \times 2x = \tfrac{1}{2} \times 5,$$

so $\qquad x = 2\tfrac{1}{2}.$

Answer 21

To find the identity element for a given operation, look at the operation table. If there is an element down the left-hand side which, when combined with the elements along the top, results in a row identical with the column headings, this is the identity element for this operation, provided that the same thing works for the column it heads, for example,

*	a	b	c
a	b	c	a
b	c	a	b
c	a	b	c

Here, c is the identity element since under the operation * it leaves the row $a\ b\ c$

a

unchanged, and the column b unchanged too.

c

To find the inverse of a, notice that $a * b = c$ so b is the inverse of a. Similarly $b * a = c$ so a is the inverse of b, and $c * c = c$, so c is its own inverse, that is c is self-inverse.

Some sets do not have identity elements under certain operations, and so there are no inverse elements either and equations cannot be solved in these sets.

See identity and inverse, Chapter 11, *SMP Book 3*.

Answer 22

Panorama takes place at 8.10 p.m.

Inverse: At 8.10 p.m. the programme on BBC1 is Panorama.

Answer 23

(a)   A function — since each man has only one head size.

(b)   Not a function, as a counting number may have several factors, for example, the factors of 12 are 1, 2, 3, 4, 6, 12.

(c)   A function, as a number can have only one 'largest' factor.

(d)   Not a function, for a woman may have more than one daughter.

See Chapter 8, *SMP Book 2*.

Chapter 7

Answer 24

Finding out about geometry.

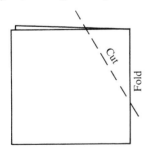
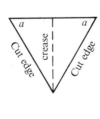

The shape is a triangle, two sides (the cut edges) are equal, the angles marked *a* are equal. The fold line is a line of symmetry and bisects the third side (the base) at right angles; it also bisects the area of the triangle.

Answer 25

A rhombus.

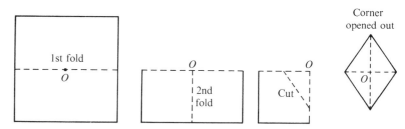

All sides are equal; opposite angles of the figure are equal; the corner angles are bisected by the fold lines; the fold lines are lines of symmetry. The fold lines (diagonals) are probably not of equal length — if they are the figure is a square and the cutting line was at 45° to the fold line. The fold lines bisect each other at right angles.

For paper folding see Exercise C, Chapter 13, *SMP Book 1*.

Answer 26

The walking insect.

No matter how many sides the polygon has, the sum of the angles is always 4 right angles. If the figure has a re-entrant angle, care must be taken to distinguish between positive (anti-clockwise) and negative (clockwise) turning.

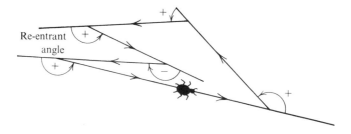

Answer 27

A translation is specified by a direction and a distance.

Answer 28

It is also true that **QH** = **T**.

Answer 29

If the two rotations of the wardrobe result in a translation the angle of **R**₁ + the angle of **R**₂ = 0° (remember to take into account the positive or negative direction of each rotation).

Answer 30

The image of a point on the mirror line is the point itself.

Answer 31

If one figure is the mirror image of the other, the perpendiculars from X and X' onto the axis meet the axis at the same point, so it is true that a reflection is a screw with a throw of zero length. (Compare this with the notion that a translation is a rotation with its centre 'at infinity'. SMP)

Fig. 7.22 contains examples of translation, rotation, reflection and screw transformations.

Chapter 8

Answer 32

Every point is a 2-node — unless it is a junction forming some other order of node. It does not need to have a 'kink' in it. Thus A, B and C are all 2-nodes.

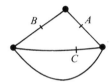

Answer 33

A topological transformation cannot change the number of arcs bounding a region. The diagrams in Fig. 8.3(c) are not equivalent, since the left-hand one has a region of order 4 while the right-hand one does not.

Answer 34

Figs. 8.4 and 8.3(c) compared.

Each figure has 4 regions.

Fig. 8.4: numbers of arcs bounding the regions are:

$$2, 4, 2, 4; \quad 1, 2, 1, 6; \quad 1, 3, 2, 5.$$
$$\text{Fig. 8.3}(c); \quad 4, 2, 4, 2; \quad 3, 3, 3, 3.$$

In each case the arcs bounding the outside region are given last.

If the left-hand diagrams were made up as a network of wires they could be shown to be equivalent in three dimensions.

For topology, see Chapter 1, *SMP Book 2*.

Answer 35

AB is the inverse of **BA** and vice versa, since they add up to a zero (stay put) vector which is plainly the identity for vector addition.

Answer 36

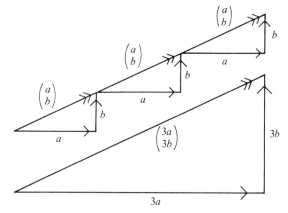

$$3\begin{pmatrix} a \\ b \end{pmatrix} = \begin{pmatrix} 3a \\ 3b \end{pmatrix}.$$

Answer 37

(a) $\begin{pmatrix} 5 \\ 5 \end{pmatrix}$;

(b) $\begin{pmatrix} 15 \\ -15 \end{pmatrix} + \begin{pmatrix} 16 \\ -8 \end{pmatrix} = \begin{pmatrix} 31 \\ -23 \end{pmatrix}$;

(c) $\begin{pmatrix} x \\ y \end{pmatrix} + \begin{pmatrix} 3 \\ 2 \end{pmatrix} = \begin{pmatrix} 7 \\ 6 \end{pmatrix} \Rightarrow \begin{pmatrix} x \\ y \end{pmatrix} = \begin{pmatrix} 7 \\ 6 \end{pmatrix} - \begin{pmatrix} 3 \\ 2 \end{pmatrix} = \begin{pmatrix} 7 \\ 6 \end{pmatrix} + \begin{pmatrix} -3 \\ -2 \end{pmatrix} = \begin{pmatrix} 4 \\ 4 \end{pmatrix}$.

(d) Rough figure – not drawn to scale.

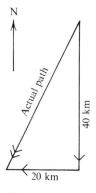

From large scale drawing or calculation using Pythagoras's theorem and trigonometry, bird's speed is approx. 44.8 km/h and its direction is South $26\frac{1}{2}°$ West, or on a bearing of 206.5°.

Answer 38

(a) $\begin{pmatrix} 5 \\ 6 \\ 7 \end{pmatrix} + \begin{pmatrix} 2 \\ 1 \\ 4 \end{pmatrix} + \begin{pmatrix} 0 \\ -1 \\ 0 \end{pmatrix} = \begin{pmatrix} 7 \\ 6 \\ 11 \end{pmatrix}$;

(b) $\begin{pmatrix} 30 \\ 35 \\ 40 \end{pmatrix}$;

(c) Can $a\begin{pmatrix} 10 \\ 8 \\ -1 \end{pmatrix} = \begin{pmatrix} 1200 \\ 900 \\ -150 \end{pmatrix}$?

It is not possible to find a value for 'a' such that $10a = 1200$ and $8a = 900$ and $-a = -150$ so the plane is not on the right path.

Answer 39

To map $ABCD \to EFGH$, reflect in the plane through the x- and y-axes, that is, in $z = 0$.

To map $ABCD \to FEHG$ rotate by $180°$ about $Y'OY$.

Answer 40

$$OG = \begin{pmatrix} 1 \\ 1 \\ -1 \end{pmatrix}, \quad OB = \begin{pmatrix} 1 \\ -1 \\ 1 \end{pmatrix}, \quad \tfrac{1}{2}(OG + OB) = \tfrac{1}{2} \left[\begin{pmatrix} 1 \\ 1 \\ -1 \end{pmatrix} + \begin{pmatrix} 1 \\ -1 \\ 1 \end{pmatrix} \right]$$

$$= \tfrac{1}{2} \begin{pmatrix} 2 \\ 0 \\ 0 \end{pmatrix}$$

$$= \begin{pmatrix} 1 \\ 0 \\ 0 \end{pmatrix}$$

$$= OX.$$

The mid-point of **DX**.

$$OD = \begin{pmatrix} -1 \\ 1 \\ 1 \end{pmatrix}, \quad OX = \begin{pmatrix} 1 \\ 0 \\ 0 \end{pmatrix}.$$

Mid-point of **DX** $= \tfrac{1}{2}(OD + OX)$

$$= \tfrac{1}{2} \left[\begin{pmatrix} -1 \\ 1 \\ 1 \end{pmatrix} + \begin{pmatrix} 1 \\ 0 \\ 0 \end{pmatrix} \right]$$

$$= \tfrac{1}{2} \begin{pmatrix} -1 \\ 1 \\ 1 \end{pmatrix}$$

$$= \begin{pmatrix} -\tfrac{1}{2} \\ \tfrac{1}{2} \\ \tfrac{1}{2} \end{pmatrix}.$$

For work on vectors see Chapter 7 of *SMP Book 2*.

Chapter 9

Answer 41

If lengths are doubled, areas are 4 times as great.

Similarly, if lengths are trebled, areas are increased 9-fold.

   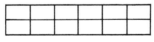

3 units of area   12 units of area

Answer 42

Strictly speaking a kilogram is a mass, but it is common practice to talk of weight. In the text, p. 38, we followed convention. Here we use the proper term.

Mass of 8 crates of soap is 240 kilograms.

Mass of 15 crates of beans is 750 kilograms.

Total mass 990 kilograms.

The load is safe.

Answer 43

Problem 1

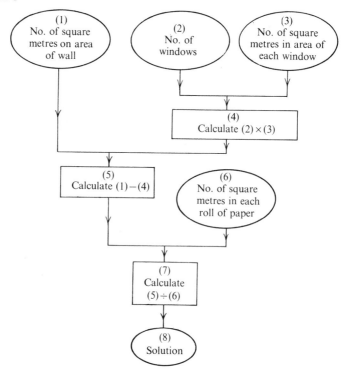

Problem 2

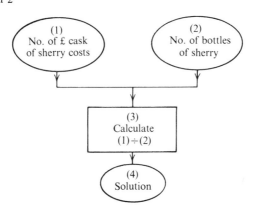

Problem 1 has the structure of Fig. 9.4.

Note that Problem 2 has superfluous information (the capacity of the cask). This 'distractor' is not an accident. Real-life problems are like this.

See also Section 5, Chapter 14, *SMP Book 1*.

Chapter 10

Answer 44

Arithmetic mean $103\frac{2}{3}$ (no, it does not matter that it is fractional) mode 121, median 104. Note that the mode is not a useful average for such a set of figures.

Answer 45

In Fig. 10.1, the left-hand side illustrates the fact that exports between the Six were just over five times as great in 1970 as in 1958. The diagram shows a tree roughly five times as high and five times as wide which thus covers twenty-five times the area of paper giving an exaggerated impression of the increase.

Chapter 12

Answer 46

$$(2 \quad 1 \quad 0) \begin{pmatrix} 17 \\ 11 \\ 2 \end{pmatrix} = 2 \times 17 + 1 \times 11 + 0 \times 2 = 45$$

$$(4 \quad 3 \quad 15) \begin{pmatrix} 17 \\ 11 \\ 2 \end{pmatrix} = 4 \times 17 + 3 \times 11 + 15 \times 2 = 131$$

Answer 47

The number of columns in the matrix on the left must equal the number of rows in the matrix on the right.

Answer 48

(a) $\begin{pmatrix} 1 & 2 \\ 2 & 3 \end{pmatrix}\begin{pmatrix} 2 \\ 4 \end{pmatrix} = \begin{pmatrix} 10 \\ 16 \end{pmatrix}$;   (b)   not possible;

(c) $\begin{pmatrix} 7 & 10 \\ 12 & 4 \end{pmatrix}$;   (d) $\begin{pmatrix} 27 & 23 & 11 \\ 1 & 5 & 15 \end{pmatrix}$;   (e)   (8   13);

(f)   not possible;   (g) $\begin{pmatrix} 4 & 5 \\ 6 & 7 \end{pmatrix}$.

Answer 49

$\mathbf{A} + \mathbf{B} = \mathbf{B} + \mathbf{A} = \begin{pmatrix} 3 & 7 \\ 5 & 5 \end{pmatrix}$.

It is always true that matrix addition is commutative.

Answer 50

(a) $\begin{pmatrix} 1 & 0 \\ 0 & -1 \end{pmatrix}$ represents a reflection in the line $y = 0$, the x-axis;

(b) $\begin{pmatrix} -1 & 0 \\ 0 & -1 \end{pmatrix}$ represents a half-turn about the origin.

See Chapter 3 of *SMP Book 3*.

Answer 51

$$\mathbf{N} = \begin{array}{c} \\ 1 \\ 2 \\ 3 \\ 4 \\ 5 \end{array} \begin{array}{cccccccc} a & b & c & d & e & f & g & h \\ \left(\begin{array}{cccccccc} 1 & 1 & 1 & 0 & 0 & 0 & 0 & 1 \\ 0 & 0 & 0 & 1 & 1 & 0 & 0 & 1 \\ 0 & 0 & 1 & 0 & 1 & 1 & 0 & 0 \\ 0 & 1 & 0 & 0 & 0 & 1 & 1 & 0 \\ 1 & 0 & 0 & 1 & 0 & 0 & 1 & 0 \end{array} \right) \end{array}$$

$$\mathbf{MN} = \begin{pmatrix} 2 & 2 & 1 & 1 & 0 & 1 & 2 & 1 \\ 1 & 1 & 1 & 2 & 2 & 2 & 2 & 1 \\ 2 & 1 & 1 & 2 & 1 & 0 & 1 & 2 \\ 1 & 1 & 2 & 1 & 2 & 1 & 0 & 2 \\ 1 & 2 & 2 & 0 & 1 & 2 & 1 & 1 \end{pmatrix}$$

This tells us the number of regions of which the node in a given row and the arc in the given column both form part of the boundary.

Nodes onto arcs:

$$\mathbf{P} = \begin{array}{c} \\ A \\ B \\ C \\ D \\ E \end{array} \begin{array}{cccccccc} a & b & c & d & e & f & g & h \\ \left(\begin{array}{cccccccc} 1 & 1 & 0 & 0 & 0 & 0 & 1 & 0 \\ 0 & 0 & 0 & 1 & 1 & 1 & 1 & 0 \\ 1 & 0 & 0 & 1 & 0 & 0 & 0 & 1 \\ 0 & 0 & 1 & 0 & 1 & 0 & 0 & 1 \\ 0 & 1 & 1 & 0 & 0 & 1 & 0 & 0 \end{array} \right) \end{array}$$

The pattern of ones in **P** is the same as the pattern of twos in **MN** (every arc ends in two nodes).

The zeros in **P** correspond either to 1 in **MN** (one region in common) or a 0 (none).

For work on networks see Chapter 6, *SMP Book 3*, and *Some Lessons in Mathematics* by T.J. Fletcher (C.U.P. 1964) Chapter 12.

Glossary on the British education system

Note. Children in Great Britain normally enter state primary schools at the age of five. When they are eleven (twelve in Scotland) they transfer to a secondary school, of which there are several different types. They remain at school until they are sixteen. If they decide to continue their education beyond the compulsory age they normally sit either the General Certificate of Education (GCE) or the Certificate of Secondary Education (CSE) before leaving the secondary school. Independent schools provide an alternative to state secondary education for those who can afford to pay the fees required or who get assistance from charitable or state sources. These include the public schools, many of them famous, richly endowed and with long traditions. Children normally enter these from fee-paying preparatory schools at the age of thirteen.

A-level, see *GCE*.

Colleges of Education: formerly called training colleges, these provide professional education, by means of three- or four-year courses, for the majority of the teaching profession.

Comprehensive schools: secondary schools that cater for the full ability-range. Most are co-educational, but there are some single-sex comprehensive schools.

Course work: work carried out as a normal part of the curriculum. It is often used for assessment purposes in CSE and GCE examinations.

CSE, Certificate of Secondary Education: a certificate of attainment for less academic pupils than those taking GCE O-level in secondary schools. Its scope is largely controlled by the teachers, especially under Mode 3, in which form the syllabus is planned, and tests are set and marked within the schools, subject to external moderation.

Department of Education: section of a university devoted to post-graduate training of teachers.

Department of Education and Science: the government ministry responsible for education.

Eleven-plus: the name loosely applied to any method of selecting children at the end of their primary schooling for different types of secondary school. Tests normally included an assessment of intelligence, and of ability in English and mathematics.

Examinations: internal examinations are set annually or more frequently in most secondary schools as a means of internal assessment; external examinations are the GCE and CSE. Amalgamation of GCE O-level and CSE is currently under discussion.

Examining Boards: the GCE and CSE examinations are administered by eight examining boards which have different syllabuses and set their own examinations. Control over standards is exercised by the Schools Council for the Curriculum and Examinations.

GCE, the General Certificate of Education: awarded in England and Wales by the eight examining boards. The Ordinary level (O-level) examination is taken at about sixteen and a good performance is essential for admission to many careers. The Advanced level (A-level), taken at about eighteen after two years in the sixth form, usually in three (but sometimes only in two) related subjects, provides the basis for admission to universities and colleges.

Grammar schools: selective secondary schools catering for more academically able pupils, and providing courses leading to the universities and colleges of education.

HMI, Her Majesty's Inspectors of Schools: responsible for advising the schools, for maintaining standards, and for providing a link between local education authorities and the government ministries responsible for education.

Honours degrees: British universities usually award degrees at honours and at general levels, the latter requiring a broader spread of studies but less depth. General degree courses are declining in importance. They are being replaced by general honours degrees in which two or three subjects are taken together in a combined honours course. Prospective teachers strive for an honours degree rather than a general degree because it is at present rewarded by a special allowance in their salary.

LEA, Local Education Authority: responsible for ensuring that adequate provision is made in their area for primary and secondary education. They provide grants for pupils seeking higher education and facilities for further education in their area.

Middle school: (*a*) the term loosely applied to third- and fourth-year (thirteen- to fifteen-year-old) pupils in a secondary school; (*b*) a school catering for the age-range nine to thirteen or fourteen. See *Upper School*.

Mixed ability: there is a move in many comprehensive schools towards teaching pupils in groups covering the whole ability range, rather than streaming them.

Non-specialist sixth form: sixth-form pupils normally follow a highly specialised curriculum based on two or three subjects, all related to 'arts' or 'sciences'. There have been repeated attempts to mitigate this evil by developing a range of 'general studies' courses to cover other subjects.

Nuffield Projects: the Nuffield Foundation (Nuffield Lodge, Regent's Park, London NW1) has initiated a number of curriculum development projects in schools, notably in science, mathematics and modern languages.

O-level, the Ordinary level of the GCE examination: normally 8 or 9 subjects are examined at the end of the fifth year in secondary schools, that is, at the age of sixteen. A pass in 4 or 5 subjects is often the minimum qualification for entering academic courses in sixth forms. Passes in specific subjects at O-level may be required as a qualification for further education or for professional qualifications of many kinds.

Schools Council for the Curriculum and Examinations: set up in 1964 to bring together representatives of all parts of the education system. The council sets up research teams, curriculum development projects, publishes working papers and reports, and helps to found local Teachers' Centres.

Secondary modern schools: selective schools catering for non-grammar secondary pupils. They were intended to have 'parity of esteem' with grammar schools, but the social effect of selection at eleven-plus has been to devalue the modern schools. Many have now been reorganised as comprehensive schools.

Setting: many secondary schools 'set' their pupils according to ability in different subjects. Sets thus become the teaching units rather than classes or forms. See also *streaming*.

Sixth form: the top age-range in grammar and comprehensive schools, catering for pupils aged from sixteen to eighteen or nineteen. It is usually organised in two year-groups, lower (or first-year) and upper (or second-year). Big schools may have a third-year sixth as well.

Streaming: a method of sorting secondary (and occasionally primary) pupils into teaching groups according to academic ability. It has been widely criticised but remains the most usual method in secondary schools.

Teachers' Centres: local centres set up under the aegis of the Schools Council to encourage discussion and liaison among practising teachers.

Upper School: a school for children aged thirteen or fourteen and up. In some areas the 'two-tier' system (primary, secondary) has been replaced by 'three-tier' (primary, middle, upper).

Index